THE ANNEALING ALGORITHM

THE KLUWER INTERNATIONAL SERIES IN ENGINEERING AND COMPUTER SCIENCE

VLSI, COMPUTER ARCHITECTURE AND DIGITAL SIGNAL PROCESSING

Consulting Editor
Jonathan Allen

Other books in the series:

Logic Minimization Algorithms for VLSI Synthesis. R.K. Brayton, G.D. Hachtel, C.T. McMullen, and A.L. Sangiovanni-Vincentelli. ISBN 0-89838-164-9.
Adaptive Filters: Structures, Algorithms, and Applications. M.L. Honig and D.G. Messerschmitt. ISBN 0-89838-163-0.
Introduction to VLSI Silicon Devices: Physics, Technology and Characterization. B. El-Kareh and R.J. Bombard. ISBN 0-89838-210-6.
Latchup in CMOS Technology: The Problem and Its Cure. R.R. Troutman. ISBN 0-89838-215-7.
Digital CMOS Circuit Design. M. Annaratone. ISBN 0-89838-224-6.
The Bounding Approach to VLSI Circuit Simulation. C.A. Zukowski. ISBN 0-89838-176-2.
Multi-Level Simulation for VLSI Design. D.D. Hill and D.R. Coelho. ISBN 0-89838-184-3.
Relaxation Techniques for the Simulation of VLSI Circuits. J. White and A. Sangiovanni-Vincentelli. ISBN 0-89838-186-X.
VLSI CAD Tools and Applications. W. Fichtner and M. Morf, editors. ISBN 0-89838-193-2.
A VLSI Architecture for Concurrent Data Structures. W.J. Dally. ISBN 0-89838-235-1.
Yield Simulation for Integrated Circuits. D.M.H. Walker. ISBN 0-89838-244-0.
VLSI Specification, Verification and Synthesis. G. Birtwistle and P.A. Subrahmanyam. ISBN 0-89838-246-7.
Fundamentals of Computer-Aided Circuit Simulation. W.J. McCalla. ISBN 0-89838-248-3.
Serial Data Computation. S.G. Smith and P.B. Denyer. ISBN 0-89838-253-X.
Phonologic Parsing in Speech Recognition. K.W. Church. ISBN 0-89838-250-5.
Simulated Annealing for VLSI Design. D.F. Wong, H.W. Leong, and C.L. Liu. ISBN 0-89838-256-4.
Polycrystalline Silicon for Integrated Circuit Applications. T. Kamins. ISBN 0-89838-259-9.
FET Modeling for Circuit Simulation. D. Divekar. ISBN 0-89838-264-5.
VLSI Placement and Global Routing Using Simulated Annealing. C. Sechen. ISBN 0-89838-281-5.
Adaptive Filters and Equalisers. B. Mulgrew, C.F.N. Cowan. ISBN 0-89838-285-8.
Computer-Aided Design and VLSI Device Development, Second Edition. K.M. Cham, S-Y. Oh, J.L. Moll, K. Lee, P. Vande Voorde, D. Chin. ISBN: 0-89838-277-7.
Automatic Speech Recognition. K-F. Lee. ISBN 0-89838-296-3.
Speech Time-Frequency Representations. M.D. Riley. ISBN 0-89838-298-X
A Systolic Array Optimizing Compiler. M.S. Lam. ISBN: 0-89838-300-5.
Algorithms and Techniques for VLSI Layout Synthesis. D. Hill, D. Shugard, J. Fishburn, K. Keutzer. ISBN: 0-89838-301-3.
Switch-Level Timing Simulation of MOS VLSI Circuits. V.B. Rao, D.V. Overhauser, T.N. Trick, I.N. Hajj. ISBN 0-89838-302-1
VLSI for Artificial Intelligence. J.G. Delgado-Frias, W.R. Moore, Editors. ISBN 0-7923-9000-8.
Wafer Level Integrated Systems: Implementation Issues. S.K. Tewskbury. ISBN 0-7923-9006-7.

THE ANNEALING ALGORITHM

by

R.H.J.M. Otten
Delft University of Technology

and

L.P.P.P. van Ginneken
Eindhoven University of Technology

KLUWER ACADEMIC PUBLISHERS
Boston/Dordrecht/London

Distributors for North America:
Kluwer Academic Publishers
101 Philip Drive
Assinippi Park
Norwell, Massachusetts 02061 USA

Distributors for all other countries:
Kluwer Academic Publishers Group
Distribution Centre
Post Office Box 322
3300 AH Dordrecht, THE NETHERLANDS

Library of Congress Cataloging-in-Publication Data

Otten, R. H. J. M.
The annealing algorithm / by R.H.J.M. Otten and L.P.P.P. van Ginneken.
p. cm.— (The Kluwer international series in engineering and computer science ; 72. VLSI, computer architecture, and digital signal processing)
Bibliography: p.
Includes index.
ISBN 0-7923-9022-9
1. Simulated annealing (Mathematics) 2. Mathematical optimization. I. Ginneken, L. P. P. P. van. II. Title. III. Series: Kluwer international series in engineering and computer science ; SECS 72. IV. Series: Kluwer international series in engineering and computer science. VLSI, computer architecture, and digital signal processing.
QA402.5.088 1989
519.3—dc20 89-8042
CIP

Printed in the United States of America

CONTENTS

PREFACE ix

1 THE ANNEALING ALGORITHM: A PREVIEW 1
1.1 Combinatorial optimization 1
1.2 Moves and local minima 5
1.3 Hill climbing 9
1.4 Simulated annealing 14
1.5 Applications 17
1.6 Mathematical model 19
1.7 Discussion 20

2 PRELIMINARIES FROM MATRIX THEORY 21
2.1 Matrices. Notation and basic properties 21
2.2 Pseudo-diagonal normal forms 25
2.3 Norms and limits of matrices 34
2.4 Quadratic forms 42
2.5 Discussion 46

3 CHAINS 47
3.1 Terminology 48
3.2 Linear arrangement, an example 50
3.3 The chain limit theorem 52
3.4 Reversible chains 59
3.5 Discussion 63

4 CHAIN STATISTICS 65
4.1 Density Functions . . . 66
4.2 Expected values . . . 68
4.3 Sampling . . . 71
4.4 Maximum likelyhood densities . . . 73
4.5 Aggregate functions . . . 74
4.6 Discussion . . . 78

5 ANNEALING CHAINS 79
5.1 Towards low scores . . . 80
5.2 Maximal accessibility . . . 86
5.3 The acceptance function . . . 89
5.4 Properties of annealing chains . . . 91
5.5 Discussion . . . 93

6 SAMPLES FROM NORMAL DISTRIBUTIONS 95
6.1 Characteristic functions . . . 95
6.2 Quadratic forms and characteristic functions . . . 100
6.3 Sampling distributions . . . 106
6.4 Asymptotic properties of sampling distributions . . . 112
6.5 Discussion . . . 113

7 SCORE DENSITIES 115
7.1 The density of states . . . 115
7.2 Weak control . . . 117
7.3 Strong control . . . 119
7.4 Three parameter aggregates . . . 122
7.5 Discussion . . . 126

8 THE CONTROL PARAMETER 127
8.1 Initialization . . . 128
8.2 Decrements in the control parameter . . . 131
8.3 A stop criterion . . . 135
8.4 Proper convergence . . . 138
8.5 Discussion . . . 139

9 FINITE-TIME BEHAVIOR OF THE ANNEALING ALGORITHM 141
9.1 Rate of convergence of chains 142
9.2 Minimum number of iterations 144
9.3 Finite-time optimal schedules 148
9.4 Discussion 150

10 THE STRUCTURE OF THE STATE SPACE 153
10.1 Chain convergence 154
10.2 The topography of the state space. 156
10.3 The set of moves 159
10.4 Global convergence 164
10.5 Discussion 165

11 IMPLEMENTATION ASPECTS 167
11.1 An implementation 167
11.2 The selection function 175
11.3 Other speed-up methods 176

REFERENCES 179

INDEX 195

PREFACE

The goal of the research out of which this monograph grew, was to make annealing as much as possible a general purpose optimization routine. At first glance this may seem a straight-forward task, for the formulation of its concept suggests applicability to any combinatorial optimization problem. All that is needed to run annealing on such a problem is a unique representation for each configuration, a procedure for measuring its quality, and a neighbor relation. Much more is needed however for obtaining acceptable results consistently in a reasonably short time. It is even doubtful whether the problem can be formulated such that annealing becomes an adequate approach for all instances of an optimization problem. Questions such as what is the best formulation for a given instance, and how should the process be controlled, have to be answered. Although much progress has been made in the years after the introduction of the concept into the field of combinatorial optimization in 1981, some important questions still do not have a definitive answer.

In this book the reader will find the foundations of annealing in a self-contained and consistent presentation. Although the physical analogue from which the concept emanated is mentioned in the first chapter, all theory is developed within the framework of markov chains. To achieve a high degree of instance independence adaptive strategies are introduced. Much emphasis is on these topics throughout the book, and in the last chapter a pascal implementation of these strategies is included. Every fact used in that implementation is proven in the preceding chapters. The book is therefore at the same time an introduction into annealing and its applications, a compendium for the theoretical background of annealing, a basis for further research, and a report on the progress made in developing a multi-purpose annealing routine. A recipe for state space construction cannot be given. Even an efficient analysis of the space regarding its adequacy for annealing has not yet appeared. But the aspects of state space construction are thoroughly discussed and illustrated.

Annealing may finally not end up being a general purpose optimization routine, but it will certainly remain a useful tool. It is easy to implement for a specific application (and hopefully this book makes access to this method even easier), and such an implementation is quite insensitive to the exact nature of the object function if only reasonable solutions are required. This is important, for example when developing a large software system, where representative solutions are needed for trying other parts of the systems, and experiments with different object functions are desirable before undertaking the much more intensive task of implementing a deterministic optimization routine specific for a certain object function. In practice, annealing has already proven its usefulness, and several gigantic design automation programs have been outperformed by the conceptually simple annealing procedure.

Most of the original research reported here has been done at the Thomas J. Watson Research Center of the IBM Corporation in Yorktown Heights, NY. That research center was also the place where annealing as a combinatorial optimization tool was conceived. Especially, Dan Gelatt's efforts to search for possible applications provided an early exposure of the concept to the authors. Many people, also outside the IBM Company, have directly and indirectly contributed to the maturing of the ideas presented in this book. These are gratefully acknowledged. This book has been prepared for printing by *Witan Presentaties*, using the LaTeX document preparation system.

Ralph H.J.M. Otten
Lukas P.P.P van Ginneken

Delft, Summer 1988

THE ANNEALING ALGORITHM

1 THE ANNEALING ALGORITHM: A PREVIEW

1.1 Combinatorial optimization

Many problems that arise in practice are concerned with finding a good or even the best configuration or parameter set from a large set of feasible options. Feasible means that the option has to satisfy a number of rigid requirements. An example of such an *optimization problem* is to find a minimum weight cylindrical can that must hold a given quantity of liquid. Assuming the weight of the empty can to be proportional to its area, the problem can be formulated as finding the dimensions of a cylinder with a volume c, and the smallest possible perimeter:

$$\begin{array}{ll} \text{minimize} & 2\pi rh + \pi r^2 \\ \text{while} & \pi r^2 h = c \quad \text{and} \quad r > 0 \end{array}$$

The formulation consists of a function to be minimized, the *objective function*, and a number of constraints. These constraints specify the set of feasible options, in this case the parameter pairs (r,h) that represent cylinders with radius r, height h, and volume c. Feasibility requirements are not always given as explicit constraints. In this example the volume constraint could have been substituted into the objective function.

If the configurations are elements of a finite or countably infinite set, the respective problems are called *combinatorial*. An example of a combinatorial optimization problem is the *assignment problem*. Suppose there are a number of men available to do an equal number of jobs, and for each man it is known how much it costs to have him perform each of these jobs. Assign to each man a job in such a way that the total cost is as small as possible.

An abstract formulation of the same problem can be as follows: given a square matrix of positive real numbers, select a set of entries such that from each row and each column exactly one entry is selected, and the sum of the selected entries is as small as possible. The equivalence between the two formulations is clear: the rows represent the men, the columns represent the jobs, and an entry is the cost of having the job associated with its column performed by the man associated with its row. The feasibility requirement for a set of entries is that it contains exactly one entry from each row and each column. The number of feasible options for an $n \times n$-matrix is $n!$, which is a finite number, though it may be a very large number.

Problems sometimes only differ in the input data. Such problems are considered to be *instances* of the same optimization problem. For example, consider the problem in which a salesman has to visit a number of cities, and is interested in the shortest tour that passes through all cities at least once and finally brings him back to the city he started from. To find the shortest tour over a number of cities with the distances between these cities given is a problem that belongs to a class of problems with equivalent formulations, but possibly with a different number of cities and different numerical values for the distances. All these problems are instances of an optimization problem generally known as the *traveling salesman problem*.

Each instance of the traveling salesman problem is characterized by a distance matrix. For n cities, and assuming a nonsymmetric matrix, this comes to $n(n-1)$ positive real numbers. The number of possible tours, that is the number of feasible configurations, is $(n-1)!$ (divided by 2 if a sequence and its inverse are considered to be the same). This is a huge number, even for small n. For 10 cities the number of possible tours is already well over 300,000. For 20 cities we are already in the 18-digit numbers. In practical applications, an instance with $n = 20$ is still small.

Salesmen are not worrying about tours over hundreds of cities, but such numbers are certainly not unusual for holes that have to be drilled in a printed circuit board. To minimize the sum of the distances over which the board has to be moved in order to be correctly positioned under the drilling machine is nothing but another instance of the traveling salesman problem. The importance of obtaining a "good" tour for this instance is evident if its length dominates the factors that determine the time needed to finish the board.

As in the case of assignment problems, also these problems can be fully specified by a square matrix, the distance matrix, and also here a configuration is a selection of exactly one entry of each row and each column of the matrix. The objective function is the same: adding the selected entries. Assignment problems, however, are not instances of the traveling salesman problem, in spite of all this similarity! Any selection of entries satisfying the constraint that there is an entry from each row and each column in the selection is a feasible option for the assignment problem. For the traveling salesman problem, however, there is the additional constraint that the entries must be the distances between successive cities on a single tour!

Feasibility can also be achieved by a generation mechanism that only produces feasible configurations. In the traveling salesman problem, the feasible configurations are all cyclic sequences of all cities. Once such a cyclic sequence has been constructed, another one can be obtained by selecting two cities and interchange their positions in the sequence. Such an operation is called a *transposition*. Any sequence can be obtained from any other sequence by successive transpositions. It can even be arranged so that every configuration is generated exactly once (counting a sequence and its inverse as distinct configurations).

What is considered feasible is usually evident in the context from which the optimization problem emanated. The measures according to which a configuration or parameter set is considered to be *good*, however, are commonly less rigid. Partly this is due to the fact that there usually are more aspects to the quality of an option than just one. A salesman may want to take lodging facilities into account. Cooperation on related jobs may have an impact on the quality of an assignment. How to handle cases with multiple objectives is not clear. Mostly one resorts to a weighted sum of quality measures or to a lexicographic ordering of objectives, but with only few applications are the results unfailingly satisfactory.

Another factor that often obscures the quality of an option is that the optimization problem is only part of a complex task. The impact on the final result of the whole task is then difficult to predict. Consequently, there often is not more than a feeling that, when confronted with two feasible options, it can be decided which option is *better* than the other. We will assume that feasible options are partially ordered by the "better relation". We even postulate the existence of an absolute quality measure, a *score function* that always assigns a lower number to a better configuration. This is necessary, because we want the problem be formulated as the minimization of the score function over the set of feasible configurations.

An *instance* is a pair ($\mathcal{S}, \varepsilon$) where $\mathcal{S}$ is a set of feasible configurations, called *states,* and ε is a *score function,* that is a mapping $\varepsilon : \mathcal{S} \rightarrow \mathbb{R}$, or quite often $\varepsilon : \mathcal{S} \rightarrow \mathbb{R}_+$.

The state set of an instance of the traveling salesman problem consists of all possible cyclic sequences of the cities. The score of each state is the sum of all distances between neighboring cities in the associated cyclic sequence. Each assignment of jobs to men (each permutation in the abstract formulation) is a state in the assignment problem. Total cost of the assignment is the score.

The goal is to find a state s in $\mathcal{S}$ such that

$$\forall_{s' \in \mathcal{S}}[\varepsilon(s) \leq \varepsilon(s')].$$

Such a state is an *optimum state.*

The formal methods to solve the instances of a certain problem are algorithms. The existence of an algorithm for a problem makes it (theoretically at least) possible to solve it by computer. For some problems algorithms are known to exist that find the best solution of any instance of such a problem in a relatively short time. Such *efficient* algorithms solve all instances of these problems in a computation time that can be bounded by a polynomial in the size of the instance. The size of a problem can be thought of as the number of bits required to specify the instance.

For many problems, however, it is generally believed that there is no efficient algorithm. The interested reader is referred to the literature on tractability of combinatorial problems, where he can learn that the traveling salesman problem belongs to a class of *np-hard* problems. It is unlikely that efficient algorithms will ever be found for problems in that class. The implication of that fact is that even for instances of moderate size the exact solution cannot always be obtained with certainty in a reasonable amount of computation time. The assignment problem, however, can be solved in computation time bounded by a polynomial of its size. In practice the best configuration of a given problem instance is seldom required. More realistically, an acceptable configuration should be produced with reasonably limited computer resources. For problems that are at least np-hard and of a size that exhaustive generation and evaluation of configurations is out of the question, the following approach may be considered.

A generation procedure is devised that produces random configurations. If the procedure does not guarantee feasibility a test should decide whether the configuration is feasible. The score of the configuration is evaluated and compared with the lowest score encountered. If the new score is lower it replaces the other and the associated configuration is stored. When a certain time limit is exceeded the process is stopped and the last stored configuration is returned. This approach is called *blind random search.*

The process is more formally described in Algorithm 1.1, where the procedure `generate` is assumed to produce only feasible configurations, each one with equal probability. The success of this method depends on the efficiency with which the configuration generation and score evaluation are implemented, and the percentage of acceptable configurations.

Algorithm 1.1

```
clock0 := clock ;
bestscore := infinity;
WHILE maxtime > clock - clock0 DO
BEGIN cconf := generate;
      cscore := epsilon(cconf);
      IF bestscore > cscore THEN
      BEGIN bestconf := cconf;
            bestscore := cscore;
      END;
END;
```

1.2 Moves and local minima

Configurations can often be generated from other configurations by small modifications, such as transpositions in the case of the traveling salesman problem. Consecutive configurations in such a generation process are strongly correlated, and the `generate` procedure in Algorithm 1.1 should therefore not be an implementation of such a modification. This strong correlation may also be reflected in the score differences. Tours obtained from another tour by a single transposition will in general not differ much in length if the number of cities is not too small. All selected distances except for four distances are the same in both configurations.

With such a generation mechanism available we may try to find better configurations by applying some of these small modifications. The scheme is then as follows. First generate a random configuration just as in the previous approach. That configuration is stored as the current configuration. Then apply a modification and evaluate the new configuration. If the score of that configuration turns out to be lower than the score of the current configuration it will take the place of the current configuration.

The process of modifying the current configuration is continued until no improvement is possible which implies a systematic exhaustive search for all modifications of the last configuration, or until a given number of subsequent modifications have taken place without improvement. The current configuration is then compared with the best configuration encountered, and if it has a lower score it will take the place of the best configuration encountered. If the maximum allowable time is not yet used another current configuration is generated randomly, and the modification process starts again.

The exact exit condition is left open in Algorithm 1.2. A possible implementation is to exit when a certain number of configurations have been generated without a single improvement. This is what the procedures at the end realize by manipulating a global variable `i`.

It is important to notice that by introducing modifications we have added something to the problem that was not inherent to the original problem. Besides, this addition is by no means unique. We have given every configuration a neighborhood, a set of configurations that can be obtained by a single application of the (chosen!) modification procedure. Such a single application is called a *move*. The set of moves can be seen as a relation over the set of states:

$$\mu \subseteq \mathcal{S} \times \mathcal{S}.$$

$(\mathcal{S}, \mu)$ is called a *state space* . $s\mu$ is the set of *neighbors* of s. When the current configuration s in the Algorithm 1.2 is such that

$$\forall_{s' \in s\mu}[\varepsilon(s) \leq \varepsilon(s')].$$

no other configuration will become the current one before a new, completely random configuration is generated. If the number of consecutive unsuccessful modifications is limited by a count or time limit this limit will certainly be exceeded, simply because none of the neighbors will have a lower score. Systematic scanning of the neighborhood will exit the loop after examining all neighbors.

A state with the property that no neighbor has a lower score is said to be *locally minimal with respect to* μ. If the move set is not clear from the context the move set has to be mentioned when discussing local minima. This is not necessary for global optima. Global optima are determined by the original problem instance and do not depend on the move set. Local optima are determined by the chosen move set.

The described modification of blind random search is called *iterative improvement*. In relatively short execution times it produces on average much better configurations than the blind search. The scores of the configurations that are compared with the best score encountered in the blind random search are distributed according to the probability density of the scores over the whole state set. The density of these scores in the case of iterative improvement is more related to the distribution of the local minima. In general, even when all neighbors of the current configuration are scanned, this density will still not be equal to the score density of the local minima, because the probability of arriving in a local minimum is not necessarily the same for all local minima. This probability depends on the selected move set. But already on the basis of the fact that local minima instead of random configurations are produced, one may expect much lower scores among the candidates for best configuration encountered.

Experiments confirm that the best scores found by iterative improvement are lower then the best scores found by blind random search. To produce one candidate in the iterative improvement scheme might take more time however than generating one random configuration. This depends on the combinatorial objects that have to be generated. It takes for example $n-2$ transpositions and one evaluation to obtain one candidate by the blind search for an n-city traveling salesman problem. A local minimum in the state space of that traveling salesman problem with transpositions as moves has already $\frac{1}{2}(n-2)(n-1)$ neighbors, and each one has to be evaluated. The time to get into the minimum has to be added to that. Nevertheless, running the algorithms with the same, relatively short, time limit usually goes in favor of iterative improvement. Of course, the difference between the score of the best configurations returned by the two methods will ultimately get smaller on average when more time is allowed.

Algorithm 1.2

```
clock0 := clock;
bestscore := infinity;
WHILE maxtime > clock - clock0 DO
BEGIN cconf := generate;
      cscore := epsilon(cconf);
      initialize;
      exit := false;
      WHILE NOT exit DO
      BEGIN exit := adjust;
            nconf := modify(cconf);
            nscore := epsilon(nconf);
            IF nscore < cscore THEN
            BEGIN exit := readjust;
                  cconf := nconf ;
                  cscore := nscore ;
            END;
      END;
      IF bestscore > cscore THEN
      BEGIN bestconf := cconf;
            bestscore := cscore;
      END;
END;

PROCEDURE initialize;
BEGIN i := 0; END;

FUNCTION  adjust: boolean;
BEGIN i := i + 1;
      adjust :=  i > maxcount;
END;

FUNCTION readjust: boolean;
BEGIN i := 0;
      readjust :=  false;
END;
```

1.3 Hill climbing

Attempting to improve the performance of such an algorithm even more one may think about ways to escape from local minima in another way than by generating a statistically independent configuration. This means leaving the idea of only moving to neighbor configurations that have a lower score than the current configuration. Instead of comparing the score of the current configuration with the score of the newly generated one, an *acceptance function* can be evaluated on the basis of the two scores involved and a random number. If a true value is returned by the acceptance function the newly generated configuration becomes the current configuration, also when that new configuration has a higher score. In the latter case the current score increases. Thus, the algorithm has the capability to escape from local minima, if there is a non-zero probability that the acceptance function returns a true value when the current score is lower than the new score.

A suggestive name has been found for this type of algorithms. The mental picture supporting this name is the state space as a hilly landscape. States that can be reached from each other by a small number of moves are thought to be close together in this landscape. The height of each state is proportional to the score function. When tracing the subsequent moves of iterative improvement we will never climb up a hill. Once we are at the bottom of a valley no change in height will occur any more. The modification proposed in the previous paragraph, on the contrary, has a non-zero probability on an uphill move. The name given to algorithms with that feature is therefore *probabilistic hill climbing algorithms* . "Probabilistic" because a random number is involved in the acceptance decision, and "hill climbing" because the score function may increase during the execution of the inner loop.

It is difficult to say anything about the performance of the algorithm without an explicit definition of the acceptance function. Hill climbing algorithms with an acceptance function that is fixed during the whole execution of the algorithm have not been studied thoroughly, probably because the performance is not expected to compare favorably with iterative improvement.

The acceptance function does not have to remain the same during the execution of the algorithm. Its evaluation process can be influenced through some control parameters. In Algorithm 1.3 these control parameters have been collected in the vector `t`. It has been organized in such a way that the algorithm executes the inner loop with certain values for the control parameters, then updates these values and reenters the loop. This continues until a stop criterion is satisfied. There are many ways, of course, to fill out the details, such as selecting the control parameters and their role in accept, how to change the values of these parameters during the execution of the algorithm, when to stop the sequence of moves and start all over with a new random configuration, etc..

Algorithm 1.3

```
clock0 := clock;
bestscore := infinity;
WHILE maxtime > clock - clock0 DO
BEGIN cconf := generate;
      cscore := epsilon(cconf);
      stop:= false;
      WHILE NOT stop DO
      BEGIN t := update(t);
            initialize;
            exit := false;
            WHILE NOT exit DO
            BEGIN exit := adjust;
                  nconf := modify(cconf);
                  nscore := epsilon(nconf);
                  IF accept(nscore, cscore, random, t) THEN
                  BEGIN exit := readjust;
                        cconf := nconf;
                        cscore := nscore;
                  END;
            END;
            stop := stopcriterion;
      END;
      IF bestscore > cscore THEN
      BEGIN bestconf := cconf;
            bestscore := cscore;
      END;
END;
```

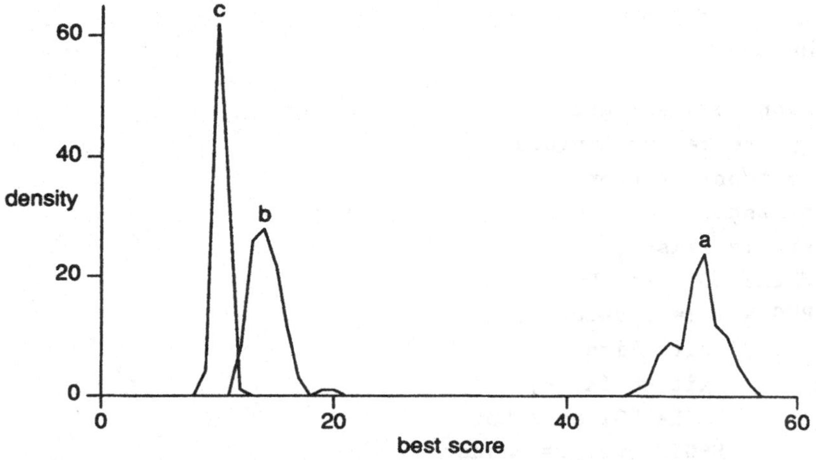

Figure 1.1: Probability densities for the configurations obtained with a) blind random search, b) iterative improvement, and c) annealing.

Recently, there is one class of probabilistic hill climbing algorithms that gets a lot of attention, namely the annealing algorithms, the subject of this book. We speak of *annealing* when the hill climbing algorithm has only one control parameter, which is positive and not increasing between the generation of statistically independent configurations by *generate*, and the acceptance function returns a true value when

$$t \ln(\texttt{random}) < \varepsilon(\texttt{cstate}) - \varepsilon(\texttt{nstate})$$

In the previous section we mentioned that the scores of the configurations that are compared with the best score encountered in the blind random search were distributed according to the probability density of the scores over the whole state set, whereas the density of these scores in the case of iterative improvement was more related to the distribution of the local minima. What about this density in the case of annealing? In figure 1.1 the densities of the scores resulting from the different approaches are plotted for blind random search, iterative improvement and some implementation of annealing. The figure does not say anything about the effort involved in obtaining the scores. It usually takes a lot more time to generate one best configuration with annealing than with iterative improvement. What the figure does suggest is that the configuration generated by annealing is likely to be better than the one generated by iterative improvement.

Algorithm 1.4

```
BEGIN cconf := generate;
      cscore := epsilon(cconf);
      bestconf := cconf;
      bestscore := cscore;
      stop:= false;
      WHILE NOT stop DO
      BEGIN t := update(t);
            initialize;
            exit := false;
            WHILE NOT exit DO
            BEGIN exit := adjust;
                  nconf := modify(cconf, t);
                  nscore := epsilon(nconf);
                  IF accept(nscore, cscore, random, t) THEN
                  BEGIN exit := readjust;
                        cconf := nconf ;
                        cscore := nscore;
                        IF bestscore > cscore THEN
                        BEGIN bestconf := cconf;
                              bestscore := cscore;
                        END;
                  END;
            END;
            stop := stopcriterion;
      END;
END;
```

The generation of completely random configurations is the same as a "modification" where the move set is the universal relation over the state set, and the acceptance function is a constant 1. So, if we are able to control the move set and the acceptance function accordingly, the outer loop has become superfluous. We therefore change Algorithm 1.3 slightly by removing the outer loop and present the result, Algorithm 1.4, as a framework for the annealing algorithm to be discussed in the remainder of this book. Another modification in Algorithm 1.4 with respect to Algorithm 1.3 is that each newly accepted configuration is compared against the best configuration generated by the algorithm. Keeping track of the best configuration encountered is therefore moved from the outer loop to the core of the algorithm. The sequence of values that t assumes during the run of the algorithm is called the *schedule.*

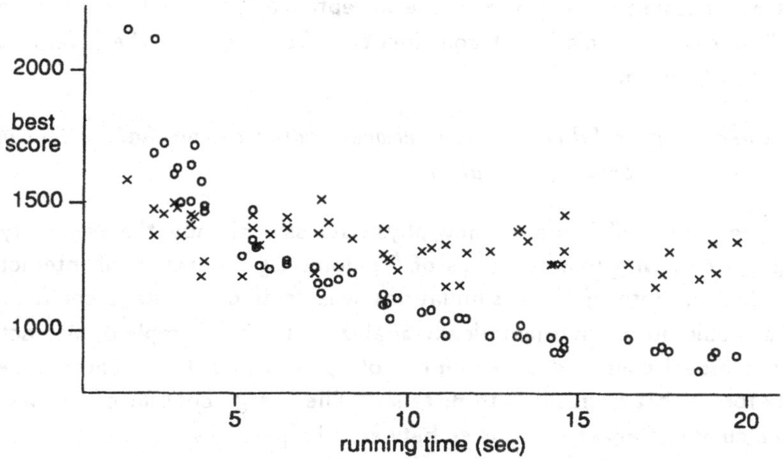

Figure 1.2: The final score for the traveling salesman problem versus the computation time: x). iterative improvement o) annealing.

To get some idea of the claims on computation time we conduct the following experiment. We take an instance of the traveling salesman problem and run an implementation of the annealing algorithm (Algorithm 1.4). After a certain amount of computation time it delivers its best score. The same amount of time is now given to an iterative improvement algorithm (Algorithm 1.2), with `maxcount` set to the square of the number of cities. The best score obtained by iterative improvement is compared with the one generated by annealing. The experiment is repeated for different computation times by slowing down the annealing implementation more and more. The results are given in figure 1.2. The figure suggests that it takes the annealing algorithm less time to obtain good solutions to the problem. For very short times both programs produce configurations of poor quality (high scores). But after some slowing down the annealing program starts to produce better solutions than the iterative improvement program. For very long execution times, not included in the experiment, the best scores of the two approaches will get closer to each other again.

1.4 Simulated annealing

The reader with some knowledge of statistical mechanics or condensed matter physics is likely to have noticed some sprinklings of those fields in the previous section, if not because of the form of the acceptance function then because of the word "annealing". This is not coincidental. Quoting from the inventors of the annealing algorithm:

There is a deep and useful connection between statistical mechanics and multivariate or combinatorial optimization.

With the computer still in its infancy physicists saw already the feasibility of simulating systems with many degrees of freedom, e.g. a system of interacting particles. The purpose of these simulations was to find the state equation of models that could not be mathematically analyzed. In the example of interacting particles the model consisted of a number of spheres in a box. These spheres interact by forces that decrease with distance. The energy content of the system could be calculated from the distances between the particles.

At each step of the simulation algorithm a new state of the system was constructed from the current state by giving a random displacement to a randomly selected particle. If the energy associated with this new state was lower than the energy of the current state the displacement was accepted, that is, the new state became the current state. If the new state had an energy higher by d joules it became with probability

$$\exp(-\frac{d}{kT})$$

the current state. This basic step can be repeated indefinitely. It is known as a *metropolis step*. The procedure is called a *metropolis loop*. It was shown that this method of generating current states led to a distribution of states in which the probability of a given state with energy e_i to be the current state was

$$\frac{\exp(-e_i/kT)}{\sum_j \exp(-e_j/kT)}.$$

This probability function is known as the boltzmann density. One of its characteristics is that for very high temperatures each state has almost equal chances of being the current state. For low temperatures only states with low energies have a high probability of being the current state. These probabilities are derived for a never ending execution of the metropolis loop. Nothing has been said about how many metropolis steps are needed for approximating these probabilities reasonably close.

Minimum energy states are called ground states in condensed matter physics. Experiments have shown that an extremely low temperature does not guarantee that the system is in its ground state or close to that state. The technique that was developed to bring a substance into one of its ground states is called *annealing*. It starts from a state in which the substance is melted. Then the temperature is slowly lowered, slowly enough to keep the system in quasi-equilibrium. When the temperature is lowered too fast, the resulting crystal may have many defects, or even lack all crystalline order.

```
PROCEDURE metropolis(t);

BEGIN initialize;
      exit := false;
      WHILE NOT exit DO
      BEGIN exit := adjust;
            nstate := displace(cstate);
            nenergy := E(nstate);
            IF random < exp ( - (nenergy - cenergy)/kt) THEN
            BEGIN exit := readjust;
                  cstate := nstate;
                  cenergy := nenergy;
            END;
      END;
END; { metropolis }
```

To apply this lesson from physics in the simulation context of the previous paragraph we have to control the variable representing the temperature. In Algorithm 1.5 the metropolis loop is therefore embedded in an outer loop. It has, in addition to the metropolis loop, a function that assigns a new value to the control parameter t, the simulation equivalent of the temperature. How many steps are executed in each metropolis loop can be controlled by the adjust functions for the exit-variable. The implications of the physical analogue are that the number of metropolis steps for a given value of t should be large enough to approximate the *equilibrium density*, that is the boltzmann density function, and the decreases in t should be small enough to keep the total number of steps small.

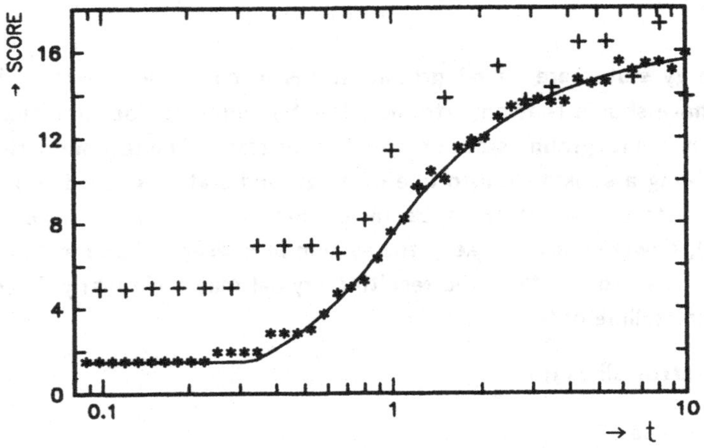

Figure 1.3: Three different annealing schedules. The full line represents a very slow schedule: long metropolis loops and small decrements in t. The asterisks represent a faster schedule: shorter metropolis loops and bigger decrements in t. The +-marks are obtained with the same number of steps in the metropolis loops, but bigger decrements in t.

In figure 1.3 we have compared the average energy of finite metropolis loops in three different runs of the algorithm. The full curve is for an extremely slow decrease of t, and many steps in the metropolis loop. The average energy follows a smooth curve and finally produces a very low value. Note that the curve is produced from the right to the left: we start with a high value of t, and slowly decrease that value! The average energies indicated by the asterisks in the figure were obtained with larger decrements in t, and shorter metropolis loops. A curve through these points would be less smooth and is clearly not monotonic, but the final energy is still very low. The results of the third run are indicated by cross signs. The decrements in t were further increased for this run. The average energies show a more unruly behavior, and the final value is far from the energy of a ground state. The figure emphasizes the importance of the `decrease` and the `adjust` functions. Long metropolis loops and small decrements in t yield low final energies, but require long computation times. Short loops and large decrements in t do not require that much computation time, but the chances on a low energy state can be small. The values of t and the number of steps have to be chosen such that a low energy is obtained with high probability without wasting too much computation time.

Algorithm 1.5

```
BEGIN cstate := generate;
      cenergy := E(cstate);
      stop:= false;
      WHILE NOT stop DO
      BEGIN t := decrease(t);
            metropolis(t);
            stop := . . . . ;
      END;
END;
```

The reader will without any difficulty discover the similarity with the annealing algorithm. A state of the system of particles corresponds with a configuration in the optimization problem. The energy of the system (divided by k, Boltzmann's constant) is the counterpart of the score function. The control parameter takes the place of temperature. The connection with combinatorial optimization was, however, not established before the early eighties, that is thirty years after the ideas for the metropolis loop were published.

1.5 Applications

The only requirement for solving a combinatorial optimization problem with annealing is that there is a set of moves that can transform one solution into another. Preferably, the scores of such solutions are close. Of course, it must be possible to find the best solution by repeated moves. So, almost any combinatorial problem can be formulated as an annealing problem. However, not all problems can be solved equally efficient. For some problems the annealing must be done so slow, that other heuristics are more efficient.

Annealing is easy to implement. A simple implementation consists of not much more than a way to change one solution into another related solution and a function that computes the score of a solution. It is advantageous that these actions can be done very fast. Almost any objective can be used. Complex objectives such as a combination of several objectives can be handled easily. The algorithm is also quite insensitive to modifications of the score function. In contrast, deterministic algorithms usually depend heavily on the score function.

Annealing has been applied to many optimization problems with varying degrees of success. In the original paper [83] a number of applications were formulated as annealing problems. Among them were placement and global wiring of integrated circuits. Placement is one of the most successful applications of annealing. A detailed implementation is given in chapter 11. Computer aided design of electronic integrated circuits is probably the first area in which simulated annealing has been used extensively. Simulated annealing has been applied to such diverse problems as folding, test pattern generation, logic minimization and compaction.

In the same paper, the traveling salesman problem was also formulated as an annealing problem. This application of annealing has been considered by many researchers. It is of interest as a typical annealing application because it is easy to implement and it offers easy comparison between various move sets and with other heuristics. However as a real application, there are heuristics for this problem that outperform annealing consistently. Other applications of annealing to graph theoretic problems include graph partitioning, graph coloring and the steiner tree problem. But annealing has been applied to problems from many other areas, such as designing error correcting codes and image restoration.

To describe an annealing application, one must describe the state space, the move set and the score function. For instance, consider the graph partitioning problem. In this problem the objective is to partition the nodes of a graph into two equally sized sets, such that the number of edges between the two sets is minimal. The state space is described by a binary vector with an entry for each node in the graph. Each entry in the vector determines in which of the two sets the node is. The sets must be balanced at all times, so the number of 1's in the vector must always be equal to the number of 0's. Since the objective is to minimize the number of edges between the two sets, this number should be the score function. The move set is not determined by the problem. However, a good move set has moves that are easy to compute and result in small score changes. For the graph partitioning problem the move set may consist of exchanging two nodes in the two sets. This is equivalent to exchanging a 1 and a 0 in the state vector. Only the edges connected to those two nodes are involved in the change of the score. This both limits the score change and makes the score update cheap to compute.

To make the annealing implementation of a problem complete, the schedule must be described. It was customary to determine the schedule by trial and error. However, the problem of finding a good schedule can be stated in general and application independent terms. By making some weak assumptions a schedule control mechanism can be devised which is problem independent, that is, which can be applied to different problems without tuning. Most of this book is devoted to developing such a schedule control.

1.6 Mathematical model

The annealing algorithm operates on a state space. At the end of the execution of a step exactly one state is the current state. The probability that a given state will be the current state depends only on its score, the score of the previous state, and the value of the control parameter t. The theoretical model for describing the sequences of current states generated by the annealing algorithm is known as a *markov chain*. The essential property is that the next state does not depend on the states that have preceded the current state. The probability that s will be the next state, given that s' is the current state is denoted by $\tau(s, s', \mathrm{t})$ and is called the *transition probability*. The transition probabilities for a certain value of t can be conveniently represented by a matrix $\mathbf{T}(\mathrm{t})$, the *transition matrix*. The transition matrix of the metropolis loop does not change from step to step, because t does not change. Markov chains with constant transition matrices are called *homogeneous*. The metropolis loop can therefore be correctly modeled by a homogeneous markov chain.

Transition probabilities for states that are not connected by a move are 0. For other pairs of distinct states the probability is determined by the probability that, given the first state, the second one is selected, and the probability that, once selected, the second state is accepted as the next state. The probability that the state does not change has to be such that the sum of all transition probabilities with that state as first state is 1, because there is always exactly one current state.

The complete markov model for the annealing algorithm is therefore

$$\tau(s, s', \mathrm{t}) = \begin{cases} \alpha(\varepsilon(s), \varepsilon(s'), \mathrm{t})\,\beta(s, s') & \text{if } s \neq s' \\ 1 - \sum_{s''} \alpha(\varepsilon(s), \varepsilon(s''), \mathrm{t})\,\beta(s, s'') & \text{if } s = s' \end{cases}$$

where α is the acceptance probability function, and β is the selection probability function.

1.7 Discussion

There are several good books on combinatorial optimization covering the material in section 1.1. A useful reference on the intractability of problems is [44]. Knowledge of that field is not a prerequisite for understanding the contents of this book, but familiarity with the notion of np-hardness helps appreciating the annealing approach. Some books on combinatorial optimization pay attention to iterative improvement, although it is understandably not a major topic. The most important aspect in the present treatment is the introduction of a neighbor relation, and the eventual creation of local minima. The name *probabilistic hill climbing* was introduced in [125], with a slightly different meaning however.

The inventors of the simulated annealing concept describe their invention in a patent application entitled *Optimization of an organization of many discrete elements* (C.D. Gelatt, Jr. et al, October 30, 1981). The patent has been issued. The first publication in the public domain is in [82]. Independently similar ideas were developed and published in [26]. Both have a basis in [103] where the metropolis loop was first described. Numerous applications of simulated annealing have appeared since the concept was spread in the world outside Thomas J.Watson Research Center of the IBM Corporation, the place where the inventors worked when developing the concept. Many, but certainly not all, of the publications describing applications are included in the bibliography.

The mathematical model for annealing has become current since 1984 [99], [125], [119]. It will also be the basis of many of the analyses in this book. The fact that we approach the problem within an abstract framework should not be seen as a denial of its roots. The intuition of physicists has laid the foundations of the method, and it is still safe to say that most practical improvements and enhancements proceeded from considering the physical analogue. The contributions from other fields were often problem specific, of pure theoretical nature, and sometimes nothing else than a new terminology or even only a new name for the method.

2 PRELIMINARIES FROM MATRIX THEORY

In this chapter we review the matrix theory used in this book. The purpose of the first section is to introduce the notation and to call to memory results derived in any core curriculum with a basic matrix theory course. The other three sections are small excursions to obtain theorems often not included in such a course. First the fact that every matrix is similar to a matrix in pseudo-diagonal normal form is established. It is the pivotal theorem in the convergence proof for annealing chains which is given in chapter 3. Another fact used in that proof is Gershgorin's theorem. Its statement and proof is at the end of the third section. That section is mainly devoted to the convergence of matrix expressions and since norms are useful concepts in that context they are introduced there. That section completes the preparation for the next chapter. The definite integral obtained in the fourth section of this chapter is not used before chapter 6. It is included in this chapter because its derivation heavily depends on matrix theory, and quadratic forms in particular.

2.1 Matrices. Notation and basic properties

Matrices are represented by boldface capitals. If the matrix $\mathbf{A}$ has m rows and n columns, it is said to have *row order* m and *column order* n. This is denoted, when convenient, as $\mathbf{A}_{m,n}$. Its ith row is indicated by $\mathbf{a}_{i,\cdot}$, and its jth columns by $\mathbf{a}_{\cdot,j}$. The entries of matrices are represented by corresponding lower case letters with subscripts indicating its row and column. They are in general numbers from the complex field $\mathbb{C}$, but mostly only from the real field $\mathbb{R}$. In the latter case we call it a real matrix. $\mathbf{A}_{n,n}$ is a *square matrix* of order n. (Column) vectors are represented by boldface lower case letters. Also when convenient, their length (i.e. the number of *components*) is given by a subscript.

Special notations are used for some special matrices, like $\mathbf{I}$ for a square matrix with ones on the diagonal and zeros elsewhere (the *identity matrix*), $\mathbf{J}$ for a matrix consisting of ones exclusively, and $\mathbf{O}$ for matrices consisting of zeros exclusively. Likewise $\mathbf{j}$ is used for an all ones vector and $\mathbf{o}$ for an all zeros vector.

The *transpose* of a matrix $\mathbf{A}$ is formed by interchanging the rows and columns, and is denoted by $\mathbf{A}^T$. A matrix $\mathbf{A}$ for which $\mathbf{A}^T = \mathbf{A}$ is called *symmetric*. The *transpose* of a vector $\mathbf{a}$ is a row vector $\mathbf{a}^T$ with the same entries. The *transpose* of a row vector $\mathbf{a}^T$ is $\mathbf{a}$. The *inner product* of two vectors of equal length is a scalar: $\mathbf{a}_n^T \mathbf{b}_n = \sum_{i=1}^n a_i b_i$ The *dyadic product* of two vectors is a matrix: $\mathbf{C}_{m,n} = \mathbf{a}_m \mathbf{b}_n^T$ with $c_{ij} = a_i b_j$. In particular $\mathbf{J}_{m,n} = \mathbf{j}_m \mathbf{j}_n^T$. If $\mathbf{a}$ and $\mathbf{b}$ have real components the inequality $(\mathbf{a}^T\mathbf{b})^2 \leq (\mathbf{a}^T\mathbf{a})(\mathbf{b}^T\mathbf{b})$ is valid.

Of course, the transpose satisfies the simple properties

$$(\mathbf{A}^T)^T = \mathbf{A}, \quad (\mathbf{A}+\mathbf{B})^T = \mathbf{A}^T + \mathbf{B}^T, \quad (\mathbf{AB})^T = \mathbf{B}^T\mathbf{A}^T$$

Herein, the plus sign indicates the usual elementwise sum, and juxtaposition indicates the usual (not the elementwise) product where the column order of the first must be the same as the row order of the second: $\mathbf{C} = \mathbf{AB}$ with $c_{ij} = \mathbf{a}_{i,\cdot}^T \mathbf{b}_{\cdot,j}$. Also powers of square matrices are defined with this product:

$$\mathbf{A}^0 = \mathbf{I}, \text{ and for all } p \in \mathbb{N},\ \mathbf{A}^p = \mathbf{A}^{p-1}\mathbf{A}.$$

The *scalar product*, $s\mathbf{A}$, is the matrix obtained by multiplying each entry with s.

The *determinant* of a square matrix $\mathbf{A}$ is defined as

$$\det(\mathbf{A}_{n,n}) = \sum_{\pi \in S_n} \operatorname{sign}(\pi) \prod_{i=1}^n a_{i\pi(i)},$$

where S_n is the set of all permutations over the n indices, and $\operatorname{sign}(\pi) = 1$ if π can be written as the product of an even number of transpositions, and -1 otherwise. The *cofactor* A_{ij} of a_{ij} is defined by $(-1)^{i+j}$ times the minor of a_{ij}, where the *minor* of a_{ij} is the value of the determinant obtained after deleting the ith row and the jth column of $\mathbf{A}$.

From these definitions it follows that

$$\det(\mathbf{A}_{n,n}) = \sum_{j=1}^{n} a_{ij} A_{ij} = \sum_{i=1}^{n} a_{ij} A_{ij}, \text{ but } \sum_{k=1}^{n} a_{ik} A_{ik} = 0, \text{for } i \neq j,$$

and also that $\det(\mathbf{A}^T) = \det(\mathbf{A})$, $\det(s\mathbf{A}_{n,n}) = s^n \det(\mathbf{A}_{n,n})$ and $\det(\mathbf{AB}) = \det(\mathbf{A}) \det(\mathbf{B})$.

A vector that is the sum of scalar multiples of other vectors is called a *linear combination* of these vectors. A set of vectors are called *linearly dependent* if **o** is a linear combination of them with not all scalars equal to zero. Otherwise these vectors are *linearly independent.* The value of a determinant is unaltered if a linear combination of some of the columns is added to any column.

A *square matrix* **A** is called *singular* if $\det(\mathbf{A}) = 0$. In that case, and only in that case, there exists a non-zero vector **a** such that $\mathbf{Aa} = \mathbf{o}$. If the determinant of a square matrix is not zero then that matrix is called *nonsingular* and in that case, and only in that case, there exists a unique matrix $\mathbf{A}^{-1}$, called the *inverse* of **A**, satisfying $\mathbf{AA}^{-1} = \mathbf{A}^{-1}\mathbf{A} = \mathbf{I}$. **A** is called *orthogonal* if $\mathbf{A}^{-1} = \mathbf{A}^T$.

Whenever the inverse exists the following properties hold:

If $\mathbf{B} = \mathbf{A}^{-1}$, then $b_{ij} = \frac{A_{ji}}{\det(A)}$.

$(c\mathbf{A})^{-1} = c^{-1} \mathbf{A}^{-1}$.

$(\mathbf{AB})^{-1} = \mathbf{B}^{-1} \mathbf{A}^{-1}$.

$\mathbf{x} = \mathbf{A}^{-1} \mathbf{b}$ is the unique solution of $\mathbf{Ax} = \mathbf{b}$.

It is also easy to verify that $(b\mathbf{I}_{n,n} + a\mathbf{J}_{n,n})^{-1} = \frac{1}{b} (\mathbf{I}_{n,n} - \frac{a}{b+an} \mathbf{J}_{n,n})$

The polynomial in λ $\phi(\lambda) = \det(\mathbf{A} - \lambda\mathbf{I})$ is called the *characteristic polynomial* of the matrix $\mathbf{A}_{n,n}$. The n roots of $\phi(\lambda)$, possibly complex numbers, are called *eigenvalues* of **A**. Eigenvalues of **A** are continuous functions of the entries of **A**. The *spectrum* of **A**, denoted by $\bar{\sigma}(\mathbf{A})$, is the family of eigenvalues of **A**. We use the word family, because not all eigenvalues will be different if $\phi(\lambda)$ has multiple roots, and we want to count repetitions. The number of roots that are equal to λ is called the *multiplicity* of λ, denoted by $m(\lambda)$.

If λ is an eigenvalue of $\mathbf{A}$, then $\mathbf{A}-\lambda\mathbf{I}$ is singular. Hence, there exists a non-zero vector $\mathbf{a}$ satisfying

$$\mathbf{A}\mathbf{a} = \lambda\mathbf{a}.$$

Such a vector is called a *right eigenvector* of $\mathbf{A}$ associated with the eigenvalue λ. *Left eigenvectors* are defined similarly. Any linear combination of eigenvectors associated with the same eigenvalue is an eigenvector.

Since the coefficient of λ^n in $\phi(\lambda)$ is $(-1)^n$, we can write $\phi(\lambda)$ in terms of its roots as

$$\phi(\lambda) = \prod_{\lambda' \in \bar{\sigma}(\mathbf{A})} (\lambda' - \lambda).$$

Setting $\lambda = 0$ in both expressions for $\phi(\lambda)$ shows that

$$\det(\mathbf{A}) = \prod_{\lambda \in \bar{\sigma}(\mathbf{A})} \lambda.$$

Similarly, matching the coefficient of λ in the two expressions gives

$$\sum_{i=1}^{n} a_{ii} = \sum_{\lambda \in \bar{\sigma}(\mathbf{A})} \lambda.$$

This last quantity is called the *trace* of the matrix.

With $\alpha \in \mathbb{R}$ we have $\det(\mathbf{A}+\alpha\mathbf{I}-\lambda\mathbf{I}) = \det(\mathbf{A}-(\lambda-\alpha)\mathbf{I})$, so that if λ is an eigenvalue of $\mathbf{A}$ then $\lambda+\alpha$ is an eigenvalue of $\mathbf{A}+\alpha\mathbf{I}$. Further, $\mathbf{A}$ and $\mathbf{A}+\alpha\mathbf{I}$ have the same eigenvectors. Also $\mathbf{A}^p$ has the same eigenvectors as $\mathbf{A}$, and if $\mathbf{a}$ is an eigenvector associated with $\lambda \in \bar{\sigma}(\mathbf{A})$, then it is also an eigenvector associated with $\lambda^p \in \bar{\sigma}(\mathbf{A}^p)$. For any $p \in \mathbb{N}_0$, $\lambda \in \bar{\sigma}(\mathbf{A})$ implies $\lambda^p \in \bar{\sigma}(\mathbf{A}^p)$.

Two matrices $\mathbf{A}$ and $\mathbf{B}$ are called *similar* if there exists a nonsingular matrix $\mathbf{S}$ such that $\mathbf{A} = \mathbf{S}\mathbf{B}\mathbf{S}^{-1}$. Similarity is an equivalence relation. If $\mathbf{A}$ and $\mathbf{B}$ are similar, then $\mathbf{A}-\lambda\mathbf{I} = \mathbf{S}(\mathbf{B}-\lambda\mathbf{I})\mathbf{S}^{-1}$ and therefore, similar matrices have the same characteristic polynomials.

The eigenvalues of a real symmetric matrix are all real. An eigenvector $\mathbf{a}$ with real entries is called *standardized* if $\mathbf{a}^T\mathbf{a} = 1$. Standardized eigenvectors of a real symmetric matrix are said to form an *orthonormal* set if the inner product of two of its elements is zero. Any symmetric matrix $\mathbf{A}_{n,n}$ can be written as

$$\mathbf{A} = \mathbf{BDB}^T = \sum_{h=1}^{n} \lambda_h \mathbf{e}^{(h)} \mathbf{e}^{(h)T},$$

where $\mathbf{D}$ is a diagonal matrix with the eigenvalues of $\mathbf{A}$ on the diagonal, $\mathbf{B}$ has the standardized eigenvectors $\mathbf{e}^{(h)}$ of $\mathbf{A}$ as columns, and these eigenvectors form an orthonormal set. $\mathbf{B}$ is therefore orthogonal. If $\mathbf{A}$ is non-singular and real symmetric, then for any integer p we have $\mathbf{A}^p = \mathbf{BD}^p\mathbf{B}^T$.

2.2 Pseudo-diagonal normal forms

We will see in later chapters that the annealing process can be represented by a series of matrix multiplications, and its statistical properties can be expressed in such terms. Properties of products and powers of matrices often follow from the pattern of zeros and non-zeros in similar matrices. Matrices with specific patterns have therefore been given suggestive adjectives. The following are among the most frequently used adjectives when dealing with matrix products.

Definition 2.1 A square matrix of which all the entries below the diagonal are zero is called an *upper triangular matrix.* When all the entries above the diagonal are zero, it is a *lower triangular matrix.* A matrix that is upper triangular as well as lower triangular is a *diagonal matrix.*

■

The identity matrix is the best known example of a diagonal matrix, and in section 2.1 we already mentioned that multiplying a matrix with $\mathbf{I}$ from the left or from the right yields the same matrix. The following two theorems show that also triangularity allows us to make strong statements about products and powers of matrices.

Theorem 2.1 If $\mathbf{A}_{n,n}$ is a triangular matrix with zeros on the diagonal, then $\mathbf{A}^k = \mathbf{O}$ for $k \geq n$.

■

Proof: We will prove by induction that if $a_{ij} = 0$ whenever $j - i + 1 > 0$ then $\mathbf{B} = \mathbf{A}^k$ satisfies

$$j - i + k > 0 \Rightarrow b_{ij} = 0.$$

Then, since $j-i$ can never be less than $1-n$, the theorem follows for lower triangular matrices. The statement is obvious for $k=1$. Suppose that for $\mathbf{C}=\mathbf{A}^{k-1}$ we have

$$j-i+k-1>0 \Rightarrow c_{ij}=0.$$

Then $b_{ij}=c_{i,.}\,\mathbf{a}_{.,j}$ where $c_{i,.}$ has all entries in the columns $i+2-k$ and up equal to zero, while $\mathbf{a}_{.,j}$ has the entries in rows 1 up to and including j equal to zero. So, c_{ij} is certainly zero when $j+1 \geq i+2-k$, or equivalently $j-i+k>0$.

qed

Theorem 2.2 The product of two upper triangular matrices is an upper triangular matrix, and the product of two lower triangular matrices is a lower triangular matrix.

■

The next theorem states that every square matrix is similar to a triangular matrix, and that the entries on the diagonal can be constrained to be the eigenvalues in any prespecified order. This theorem underlines the usefulness of the two previous theorems, because a matrix product under a similarity transformation yields the same result as multiplying the matrices after they have been transformed by the same similarity transformation.

Theorem 2.3 Every square matrix is similar to an upper triangular matrix (and to a lower triangular matrix) of which the diagonal contains the complete spectrum of that matrix in any given sequence.

■

Proof: We will prove that an arbitrary matrix $\mathbf{A}_{n,n}$ is similar to an upper triangular matrix $\mathbf{U}_{n,n}$ by induction. The statement is trivial for $n=1$. Suppose the statement is true for all matrices $\mathbf{C}_{n-1,n-1}$. Let $\mathbf{v}$ be a right eigenvector of $\mathbf{A}$ associated with the first eigenvalue ν in the given sequence.

$$\sum_{j=1}^{n} a_{ij} v_j = \nu v_i.$$

We take some nonsingular matrix $\mathbf{P}_{n,n}$ of which the first column is the vector $\mathbf{v}$. With $\mathbf{Q}$ as the inverse of $\mathbf{P}$, and $\mathbf{B}=\mathbf{QAP}$, we have

$$b_{i1}=\sum_{j=1}^{n}\sum_{k=1}^{n} q_{ij}\,a_{jk}\,p_{k1}=\sum_{j=1}^{n}\left(q_{ij}\sum_{k=1}^{n} a_{jk} v_k\right)=\nu\sum_{j=1}^{n} q_{ij}\,p_{j1}=\nu\delta_{i1}.$$

($\delta_{ij} = 1$ when $i = j$ and 0 otherwise.) So, **B** is similar to **A** and has a first column consisting of zeros except for the first entry b_{11} which is equal to ν.

$$\mathbf{B} = \left[\begin{array}{c|ccc} \nu & * & \cdots & * \\ \hline 0 & & & \\ \vdots & & \mathbf{C} & \\ 0 & & & \end{array}\right]$$

$\mathbf{C}_{n-1,n-1}$ is the matrix **B** with the first column and row stripped from it. Since

$$\det(\mathbf{B}_{n,n} - \lambda \mathbf{I}_{n,n}) = (b_{11} - \lambda)\det(\mathbf{C}_{n-1,n-1} - \lambda \mathbf{I}_{n-1,n-1}),$$

C has the same spectrum as **A**, with one occurrence of ν deleted. By our hypothesis there is a non-singular **R** such that $\mathbf{R}^{-1}\mathbf{C}\mathbf{R} = \mathbf{T}$ with **T** upper triangular and the other eigenvalues in the given sequence on the diagonal. If $\mathbf{S}_{n,n}$ is the matrix **R** expanded with a first row and column consisting of 1 followed by zeros, then we have

$$\mathbf{A} = \mathbf{P}\mathbf{B}\mathbf{P}^{-1} = \mathbf{P}\mathbf{S}\mathbf{U}\mathbf{S}^{-1}\mathbf{P}^{-1} = (\mathbf{P}\mathbf{S})\mathbf{U}(\mathbf{P}\mathbf{S})^{-1},$$

U being a matrix equal to **T** with a first row and a first column added to it. Moreover, this first column is ν followed by zeros.

$$\mathbf{U} = \left[\begin{array}{c|ccc} \nu & * & \cdots & * \\ \hline 0 & & & \\ \vdots & & \mathbf{T} & \\ 0 & & & \end{array}\right]$$

Consequently, **U** is similar to **A**, and is an upper triangular matrix with the eigenvalues of **A** on the diagonal in the given sequence.

qed

But also the values of the non-zeros do influence the result of a matrix product, of course. Many so-called normal forms of matrices are characterized by the entries on the diagonal as well as the pattern of zeros and non-zeros. The ultimate normal form is the jordan form which has the eigenvalues on the diagonal and only zeros and ones on one superdiagonal. All other entries in the jordan form are zero. The jordan form is essential in the theory of invariant subspaces. We will not be involved in that theory, and the jordan form can be avoided in this book. A weaker result, but still, like the jordan form, triangular with the eigenvalues on the diagonal, is needed however.

Before being more specific about the zeros on the non-zero side of the diagonal, we want to know when the matrix equation $\mathbf{AX} - \mathbf{XB} = \mathbf{C}$ has a solution. The motivation for this question is evident from looking at the product of two appropriately partitioned matrices. It is easy to verify the correctness of the following rule:

Theorem 2.4 If $\mathbf{A}_{m,n}$ can be partitioned into blocks $\mathbf{A}^{i,j}$ with r_i rows and s_j columns, and $\mathbf{B}_{n,p}$ can be partitioned into blocks $\mathbf{B}^{j,k}$ with s_j rows and t_k columns, then the matrix $\mathbf{C}_{m,p} = \mathbf{A}_{m,n}\mathbf{B}_{n,p}$ may be partitioned into blocks $\mathbf{C}^{i,k}$ with r_i rows and t_k columns with $\mathbf{C}^{j,k} = \sum_j \mathbf{A}^{i,j}\mathbf{B}^{j,k}$.

■

Using this rule shows that

$$\begin{pmatrix} \mathbf{I} & \mathbf{O} \\ \mathbf{X} & \mathbf{I} \end{pmatrix}\begin{pmatrix} \mathbf{I} & \mathbf{O} \\ -\mathbf{X} & \mathbf{I} \end{pmatrix} = \begin{pmatrix} \mathbf{I} & \mathbf{O} \\ \mathbf{O} & \mathbf{I} \end{pmatrix}$$

Since a lower triangular matrix can be partitioned in two by two blocks where the upper-right block is a zero submatrix, it is always similar to

$$\begin{pmatrix} \mathbf{I} & \mathbf{O} \\ \mathbf{X} & \mathbf{I} \end{pmatrix}\begin{pmatrix} \mathbf{B} & \mathbf{O} \\ \mathbf{C} & \mathbf{A} \end{pmatrix}\begin{pmatrix} \mathbf{I} & \mathbf{O} \\ -\mathbf{X} & \mathbf{I} \end{pmatrix} = \begin{pmatrix} \mathbf{B} & \mathbf{O} \\ \mathbf{XB} + \mathbf{C} - \mathbf{AX} & \mathbf{A} \end{pmatrix}$$

The lower-left block can be made $\mathbf{O}$ by choosing an appropriate $\mathbf{X}$ if such an $\mathbf{X}$ exists. Since the lower-left block is equal to $\mathbf{XB} + \mathbf{C} - \mathbf{AX}$ such an $\mathbf{X}$ only exists if the mentioned matrix equation has a solution. A detour from the path towards the normal form needed is therefore necessary. On this detour we want to find the conditions for the existence of a solution to the given matrix equation. For this investigation a matrix product other than the usual one is useful.

Definition 2.2 The matrix $\mathbf{C}_{pm,qn}$ is called the *kronecker product* of the matrices $\mathbf{A}_{p,q}$ and $\mathbf{B}_{m,n}$ if $c_{ij} = a_{dd'}\, b_{rr'}$, with $i+m = md+r$, $1 \leq r \leq m$ and $j+n = nd'+r'$, $1 \leq r' \leq n$. It is denoted by $\mathbf{A} \boxtimes \mathbf{B}$.

■

$$
\mathbf{A}_{p,q} \boxtimes \mathbf{B}_{m,n} = \left[\begin{array}{c|c|c|c}
a_{11}\mathbf{B} & a_{12}\mathbf{B} & \cdots & a_{1q}\mathbf{B} \\ \hline
a_{21}\mathbf{B} & & & \\ \hline
\vdots & & & \\ \hline
a_{p1}\mathbf{B} & & & a_{pq}\mathbf{B}
\end{array} \right]
$$

Thus, the kronecker product is a matrix with a natural partitioning of p by q blocks, each block containing a matrix $a_{ij}\mathbf{B}$.

There is a simple, but very useful rule that relates usual matrix products with kronecker products. Many identities involving kronecker products can be established quite easily by using this rule. We will use the rule not to derive such identities, but to express the eigenvalues of matrices formed by kronecker products in terms of the eigenvalues of the kronecker factors. Let us therefore state and prove that rule, and then establish the relations between the eigenvalues.

Theorem 2.5 (Stephanos' rule)

$$(\mathbf{A}_{p,q} \boxtimes \mathbf{B}_{s,t})(\mathbf{C}_{q,s} \boxtimes \mathbf{D}_{t,r}) = (\mathbf{A}_{p,q}\mathbf{C}_{q,s}) \boxtimes (\mathbf{B}_{s,t}\mathbf{D}_{t,r})$$

■

Proof: Using the natural partitioning of the kronecker product and applying definition 2.2 on the left hand side yields a matrix of p by s blocks, where block^{ij} contains $\left(\sum_{k=1}^{q} a_{ik}\, c_{kj}\right) \mathbf{B}\mathbf{D}$ This is exactly the i,j-entry of $\mathbf{A}\ \mathbf{C}$ times the matrix $\mathbf{B}\ \mathbf{D}$, which is by definition the block^{ij} of the right hand side.

qed

The first question we obtain an answer to by using this rule is: what is the spectrum of the kronecker product if we know the spectra of its factors?

Theorem 2.6 The spectrum of $\mathbf{A} \boxtimes \mathbf{B}$ consists of all pairwise products of an eigenvalue of $\mathbf{A}$ and an eigenvalue of $\mathbf{B}$:

$$\bar{\sigma}(\mathbf{A}_{n,n} \boxtimes \mathbf{B}_{m,m}) = \{\lambda\mu \mid \lambda \in \bar{\sigma}(\mathbf{A}) \wedge \mu \in \bar{\sigma}(\mathbf{B})\}$$

■

Proof: Suppose $\mathbf{Aa} = \lambda\mathbf{a}$ and $\mathbf{Bb} = \mu\mathbf{b}$ where we can view $\mathbf{a}$ and $\mathbf{b}$ as 1-column matrices. Then $\mathbf{a} \boxtimes \mathbf{b}$ is a vector of mn components. With Stephanos' rule we then obtain

$$(\mathbf{A} \boxtimes \mathbf{B})(\mathbf{a} \boxtimes \mathbf{b}) = \mathbf{Aa} \boxtimes \mathbf{Bb} = (\lambda\mathbf{a}) \boxtimes (\mu\mathbf{b}) = \lambda\mu(\mathbf{a} \boxtimes \mathbf{b}).$$

By the continuity of eigenvalues we conclude that the whole spectrum is given in the statement of the theorem.

qed

More important for our investigation into the existence of a solution for the matrix equation $\mathbf{AX} - \mathbf{XB} = \mathbf{C}$ is a matrix with the sums of the eigenvalues of two matrices in its spectrum. That matrix is sometimes called the *kronecker sum*. Actually, it is the usual matrix sum of two kronecker products. Its formula is given in the formulation of the next theorem which states the fact about its spectrum.

Theorem 2.7 The spectrum of $\mathbf{A}_{n,n} \boxtimes \mathbf{I}_{m,m} + \mathbf{I}_{n,n} \boxtimes \mathbf{B}_{m,m}$ consists of all pairwise sums of an eigenvalue of $\mathbf{A}$ and an eigenvalue of $\mathbf{B}$:

$$\bar{\sigma}(\mathbf{A}_{n,n} \boxtimes \mathbf{I}_{m,m} + \mathbf{I}_{n,n} \boxtimes \mathbf{B}_{m,m}) = \{\lambda + \mu \mid \lambda \in \bar{\sigma}(\mathbf{A}) \wedge \mu \in \bar{\sigma}(\mathbf{B})\}$$

■

Proof: Adopting the same assumptions as in the previous proof, Stephanos' rule now gives us

$$(\mathbf{A} \boxtimes \mathbf{I}_{m,m} + \mathbf{I}_{n,n} \boxtimes \mathbf{B})(\mathbf{a} \boxtimes \mathbf{b}) = (\mathbf{A} \boxtimes \mathbf{I}_{m,m})(\mathbf{a} \boxtimes \mathbf{b}) + (\mathbf{I}_{n,n} \boxtimes \mathbf{B})(\mathbf{a} \boxtimes \mathbf{b}) =$$

$$(\mathbf{Aa} \boxtimes \mathbf{b}) + (\mathbf{a} \boxtimes \mathbf{Bb}) = (\lambda\mathbf{a}) \boxtimes \mathbf{b} + \mathbf{a} \boxtimes (\mu\mathbf{b}) = (\lambda + \mu)(\mathbf{a} \boxtimes \mathbf{b}).$$

Again, this specifies the complete spectrum.

qed

We need one more identity to formulate the conditions for the existence of a solution for the matrix equation. The proof of that identity may seem a bit complicated. Like many proofs involving kronecker products it hinges on the selection of the right partitioning of the matrices. The identity is then verified by showing that corresponding blocks on both sides of the equality sign are indeed equal. It is therefore often helpful to visualize the matrix partitions throughout the proof.

Theorem 2.8 (Neudecker's identity)

Let $(\mathbf{M})^{\nabla}$ be a column vector obtained by concatenating the columns of $\mathbf{M}$, where $\mathbf{M}$ may be any matrix, then

$$(\mathbf{A}_{p,q}\mathbf{B}_{q,r}\mathbf{C}_{r,t})^{\nabla} = (\mathbf{C}^{T}_{r,t} \boxtimes \mathbf{A}_{p,q})(\mathbf{B}_{q,r})^{\nabla}.$$

■

Proof: If we partition $\mathbf{D} = \mathbf{AB}$ into columns and $\mathbf{c}_{.,j}$ into single elements, then

$$\mathbf{ABc}_{.,j} = \mathbf{Dc}_{.,j} = \sum_{i=1}^{r} \mathbf{d}_{.,i}\, c_{ij} = \sum_{i=1}^{r} (c_{ij}\,\mathbf{A})\mathbf{b}_{.,i}$$

To obtain the corresponding part of the right hand side we partition $\mathbf{C}^{T} \boxtimes \mathbf{A}$ in the usual way for kronecker products, and partition $(\mathbf{B})^{\nabla}$ into blocks of q by 1. The jth block (1 by p)

$$\left((\mathbf{c}^{T})_{j,.} \boxtimes \mathbf{A}\right)(\mathbf{B})^{\nabla} = \left((\mathbf{c}_{.,j})^{T} \boxtimes \mathbf{A}\right)(\mathbf{B})^{\nabla} = \sum_{i=1}^{r} (c_{ij}\mathbf{A})\mathbf{b}_{.,i}$$

qed

The last two theorems are all we need to complete our detour, because with those theorems available it is relatively simple to obtain the condition we want.

Theorem 2.9 The matrix equation $\mathbf{AX} - \mathbf{XB} = \mathbf{C}$ has a solution for any triple $(\mathbf{A}, \mathbf{B}, \mathbf{C})$ if $\mathbf{A}$ and $\mathbf{B}$ have no eigenvalues in common:

$$\forall_{\mathbf{A}_{n,n}} \forall_{\mathbf{B}_{m,m}} \forall_{\mathbf{C}_{n,m}} \left[\forall_{(\lambda,\mu)\in\bar{\sigma}(\mathbf{A})\times\bar{\sigma}(\mathbf{B})}[\lambda \neq \mu] \Rightarrow \exists_{\mathbf{X}_{n,m}} [\mathbf{A}_{n,n}\mathbf{X}_{n,m} - \mathbf{X}_{n,m}\mathbf{B}_{m,m} = \mathbf{C}_{n,m}]\right]$$

■

Proof: Using Neudecker's identity twice yields

$$\begin{aligned}
&(\mathbf{C}_{n,m})^{\nabla} = \\
&(\mathbf{A}_{n,n}\mathbf{X}_{n,m} - \mathbf{X}_{n,m}\mathbf{B}_{m,m})^{\nabla} = \\
&(\mathbf{A}_{n,n}\mathbf{X}_{n,m})^{\nabla} - (\mathbf{X}_{n,m}\mathbf{B}_{m,m})^{\nabla} = \\
&(\mathbf{A}_{n,n}\mathbf{X}_{n,m}\mathbf{I}_{m,m})^{\nabla} - (\mathbf{I}_{n,n}\mathbf{X}_{n,m}\mathbf{B}_{m,m})^{\nabla} = \\
&(\mathbf{I}_{m,m} \boxtimes \mathbf{A}_{n,n})(\mathbf{X}_{n,m})^{\nabla} - (\mathbf{B}^{T}_{m,m} \boxtimes \mathbf{I}_{n,n})(\mathbf{X}_{n,m})^{\nabla} = \\
&(\mathbf{I}_{m,m} \boxtimes \mathbf{A}_{n,n} - \mathbf{B}^{T}_{m,m} \boxtimes \mathbf{I}_{n,n})(\mathbf{X}_{n,m})^{\nabla}
\end{aligned}$$

Since the linear system of equations

$$(\mathbf{I}_{m,m} \boxtimes \mathbf{A}_{n,n} - \mathbf{B}^{T}_{m,m} \boxtimes \mathbf{I}_{n,n})(\mathbf{X}_{n,m})^{\nabla} = (\mathbf{C}_{n,m})^{\nabla}$$

has a (unique) solution fif

$$\det(\mathbf{I}_{m,m} \boxtimes \mathbf{A}_{n,n} - \mathbf{B}^{T}_{m,m} \boxtimes \mathbf{I}_{n,n}) = \prod_{(\lambda,\mu) \in \bar{\sigma}(A) \times \bar{\sigma}(B)} (\lambda - \mu) \neq 0\,,$$

the theorem follows.

qed

With theorem 2.3 and theorem 2.9 at hand we have all we need to complete the main line of derivation. The motivation for investigating the matrix equation $\mathbf{AX} - \mathbf{XB} = \mathbf{C}$ was to have for any given matrix a similar matrix with a 2 by 2 partitioning in which both off-diagonal blocks only have zero entries. Theorem 2.3 tells us that it is always possible to have one of these blocks equal to $\mathbf{O}$, even to have all entries on one side of the diagonal equal to 0. Theorem 2.9 tells us that it is also possible to have the other block equal to $\mathbf{O}$ if the two on-diagonal blocks do not have any eigenvalue in common. We can make the most of this procedure by creating on-diagonal blocks that have only one distinct eigenvalue. The matrix is then in a form that is called the pseudo-diagonal normal form, and, as we will see in the next chapter, the existence of that form enables us to obtain the main result of that chapter in a relatively easy way.

Definition 2.3 Let $\mathbf{m}$ be a vector whose entries are the multiplicities of the eigenvalues of a given matrix. That matrix is said to be in *pseudo-diagonal normal form* if it is triangular and it can be partitioned into blocks of m_i rows and m_j columns such that all off-diagonal blocks are $\mathbf{O}$ and the entries on the diagonal of an on-diagonal block are all equal to the corresponding eigenvalue. We denote the block that has eigenvalue λ on the diagonal by $\mathbf{B}_{\lambda}$.

■

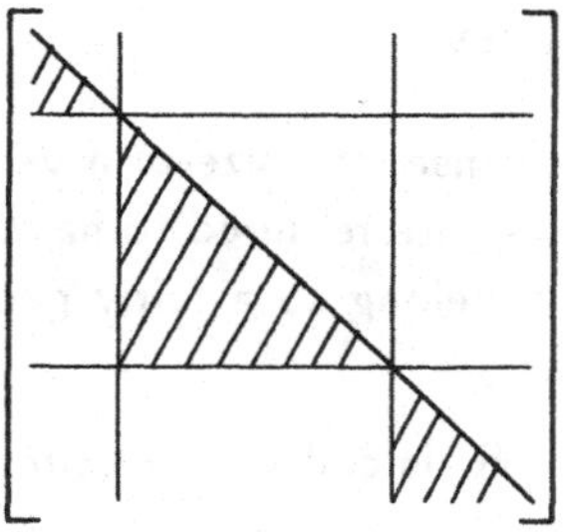

Theorem 2.10 Every square matrix is similar to a matrix in pseudo-diagonal normal form.

■

Proof: According to theorem 2.3 there exists for a given matrix $\mathbf{A}_{n,n}$ a similar matrix $\mathbf{B}$ which has the eigenvalues of $\mathbf{A}$ on the diagonal in a sequence in which the equal eigenvalues are successive. If there are at least two distinct eigenvalues we can partition $\mathbf{B}$ as

$$\mathbf{B} = \left[\begin{array}{c|c} \mathbf{E} & \mathbf{O} \\ \hline \mathbf{D} & \mathbf{F} \end{array}\right]$$

where no diagonal element of $\mathbf{E}_{m,m}$ is equal to a diagonal element of $\mathbf{F}_{n-m,n-m}$. By theorem 2.9 there exist a matrix $\mathbf{U}_{n-m,m}$ such that $\mathbf{UE} - \mathbf{FU} + \mathbf{D} = \mathbf{O}$. Consequently, the matrix $\mathbf{TBT}^{-1}$, partitioned as $\mathbf{B}$, has $\mathbf{O}_{n-m,m}$ and $\mathbf{O}_{m,n-m}$ as off-diagonal blocks, if $\mathbf{T}_{n,n}$ is the matrix

$$\mathbf{T} = \left[\begin{array}{c|c} \mathbf{I} & \mathbf{O} \\ \hline \mathbf{U} & \mathbf{I} \end{array}\right]$$

Applying the same procedure to on-diagonal blocks of $\mathbf{TBT}^{-1}$ with at least two distinct values on the diagonal will ultimately yield a matrix in pseudo-diagonal normal form.

qed

2.3 Norms and limits of matrices

Norms are used to assess, in some sense, the size of a vector or a matrix. Commonly used functions for this purpose are required to have certain properties. All the norms we will use in this book belong to a fairly general class of functions which is defined as follows:

Definition 2.4 A function $f: \mathbb{C}^n \to \mathbb{R}$ is called a *seminorm* if

positivity $\forall_{x \in \mathbb{C}^n} [f(x) \geq 0]$,

proportionality $\forall_{x \in \mathbb{C}^n} \forall_{k \in \mathbb{C}} [f(kx) = |k| f(x)]$, and

triangle inequality $\forall_{x \in \mathbb{C}^n} \forall_{y \in \mathbb{C}^n} [f(x+y) \leq f(x) + f(y)]$

are satisfied. A seminorm is called *non-trivial* if $\exists_{x \in \mathbb{C}^n} [f(x) > 0]$

■

One property that might have been expected is not explicitly mentioned in this definition. From a function assigning a number to an object that is to be interpreted as the size of that object, we expect that small changes in the object cause small variations in that number. For non-trivial seminorms we can show that this property is implicit in the definition.

Theorem 2.11 A non-trivial seminorm is a continuous function of its arguments.

■

Proof: For any $\mathbf{x} \in \mathbb{C}^n$ and any $\mathbf{d} \in \mathbb{C}^n$ we have by the triangle inequality that

$$f(\mathbf{x}+\mathbf{d}) - f(\mathbf{x}) \leq f(\mathbf{x}) + f(\mathbf{d}) - f(\mathbf{x}) = f(\mathbf{d})$$

and by the triangle inequality and the proportionality

$$f(\mathbf{x}+\mathbf{d}) - f(\mathbf{x}) = f(\mathbf{x}+\mathbf{d}) - f(\mathbf{x}+\mathbf{d}-\mathbf{d}) \geq f(\mathbf{x}+\mathbf{d}) - f(\mathbf{x}+\mathbf{d}) - f(\mathbf{d}) = -f(\mathbf{d}),$$

and therefore, $|f(\mathbf{x}+\mathbf{d}) - f(\mathbf{x})| \leq f(\mathbf{d})$.
Also by the proportionality and the triangle inequality

$$f(\mathbf{d}) = f\left(\sum_{i=1}^{n} d_i \mathbf{i}_{\cdot,i}\right) \leq \sum_{i=1}^{n} f(d_i \mathbf{i}_{\cdot,i}) =$$

$$= \sum_{i=1}^{n} |d_i| f(\mathbf{i}_{\cdot,i}) \leq \left(\max_{1 \leq j \leq n} |d_j|\right) \sum_{i=1}^{n} f(\mathbf{i}_{\cdot,i}).$$

For a non-trivial seminorm the sum $\sum_{i=1}^{n} f(\mathbf{i}_{.,i})$ cannot be zero, because of the proportionality requirement. Further, for any $\varepsilon > 0$ the set

$$\left\{ \mathbf{d} \;\middle|\; \left(\max_{1 \le j \le n} |d_j| \right) < \frac{\varepsilon}{\sum_{i=1}^{n} f(\mathbf{i}_{.,i})} \right\}$$

is not empty, and therefore,

$$\forall_{\varepsilon>0} \exists_{\mathbf{d}\in\mathbb{C}^n} \left[\,|\, f(\mathbf{x}+\mathbf{d}) - f(\mathbf{x}) \,|\, < \varepsilon \right]$$

qed

Definition 2.5 The *norm of a vector* **v**, denoted by $\|\mathbf{v}\|$, is a seminorm with the additional requirement that

$$\|\mathbf{v}\| = 0 \Leftrightarrow \mathbf{v} = \mathbf{o}.$$

■

The most common vector norms are the so-called *minkowski p-norms:*

$$\|\mathbf{v}\|_p = \left(|v_1|^p + |v_2|^p + \ldots + |v_n|^p \right)^{\frac{1}{p}}$$

where p is any natural number. In addition we have

$$\|\mathbf{v}\|_\infty = \lim_{p \to \infty} \|\mathbf{v}\|_p = \max_{1 \le j \le n} |v_j|.$$

it is quite easy to prove that the minkowski p-norms are vector norms except for the triangle inequality (which in that case is called *Minkowski's inequality*). However, for the norms that we will use in this book ($p = 1$ and ∞) the proof of that inequality is trivial as well.

Definition 2.6 The *norm of a matrix,* denoted as $\|\mathbf{A}\|$, is a seminorm which satisfies

$$\|\mathbf{A}\| = 0 \Longleftrightarrow \mathbf{A} = \mathbf{O}$$
$$\|\mathbf{AB}\| \le \|\mathbf{A}\| \, \|\mathbf{B}\|$$

A matrix norm and a vector norm that satisfy the relation

$$\|\mathbf{Ax}\| \le \|\mathbf{A}\| \, \|\mathbf{x}\|$$

for all **A** and **x** are said to be *compatible.*

■

With the interpretation of a norm as the size of its object in mind we can speculate about the properties a *consistent* matrix norm should have. Both, **v** and **Av**, are vectors, and a norm assigns a kind of length to them. It is quite natural to expect from the norm of a matrix that it tells us something about the effect of matrix multiplication on the length. In the definition of compatibility the matrix norm is required to supply an upper bound to the stretching capability of the matrix. The closer this bound is to the least upper bound, the more powerful is the norm involved. In the next theorem we see that this least upper bound is indeed a matrix norm. It is not surprising that this matrix norm is the one commonly used.

Theorem 2.12

$$\limsup_{\mathbf{x} \neq \mathbf{o}} \frac{\| \mathbf{Ax} \|}{\| \mathbf{x} \|}$$

satisfies the conditions required of a matrix norm, and is, as such, compatible with the vector norm used in the definition. This matrix norm is said to be *subordinate* to the vector norm used.

■

Proof: It can be readily verified that if $f(\mathbf{x})$ is a seminorm, then so is $g(\mathbf{x}) = f(\mathbf{Ax})$. So, $g(\mathbf{x})$ is a continuous function of its arguments. Since $\{\mathbf{x} \mid \|\mathbf{x}\| = 1\}$ is a closed bounded set no matter which vector norm is used, $g(\mathbf{x})$ has to achieve its maximum value over this set in some point say $\mathbf{y}$ where $\|\mathbf{y}\| = 1$ of course. Since for any vector $\mathbf{v}$ we can find a vector with the same direction as $\mathbf{v}$, whose norm has value 1, namely $\frac{\mathbf{v}}{\|\mathbf{v}\|}$, we have the following identities:

$$h(\mathbf{A}) = \limsup_{\mathbf{v} \neq \mathbf{o}} \frac{\|\mathbf{Av}\|}{\|\mathbf{v}\|} = \limsup_{\|\mathbf{x}\| = 1} \|\mathbf{Ax}\| = \max_{\|\mathbf{x}\| = 1} \|\mathbf{Ax}\| = \|\mathbf{Ay}\|$$

This quantity trivially satisfies positivity and proportionality, and, of course, $h(\mathbf{A})\|\mathbf{x}\| \geq \|\mathbf{Ax}\|$, for all $\mathbf{x}$. The triangle inequality is verified by noting that for $\mathbf{A} = \mathbf{B} + \mathbf{C}$

$$h(\mathbf{B} + \mathbf{C}) = \|(\mathbf{B} + \mathbf{C})\mathbf{y}\| \leq \|\mathbf{By}\| + \|\mathbf{Cy}\| \leq h(\mathbf{B}) + h(\mathbf{C}).$$

Finally, for $\mathbf{A} = \mathbf{BC}$ we have

$$h(\mathbf{A}) = \|\mathbf{BCy}\| \leq h(\mathbf{B})\|\mathbf{Cy}\| \leq h(\mathbf{B})h(\mathbf{C})\|\mathbf{y}\| = h(\mathbf{B})h(\mathbf{C}).$$

qed

It is natural at this point to investigate the matrix norms subordinate to the vector norms we have met already in this section. We will do that for the two norms that will be used in this book.

Theorem 2.13 A matrix norm subordinate to the minkowski 1-norm is the maximum sum of the moduli of the entries in a column. For $\mathbf{A}_{m,n}$: $\| \mathbf{A} \|_1 = \max_{1 \le j \le n} \sum_{i=1}^{m} | a_{ij} |$.

■

Proof: If f is a matrix norm subordinate to the minkowski 1-norm, then there is a $\mathbf{y}$ such that $\| \mathbf{y} \|_1 = 1$ and $f(\mathbf{A}) = \| \mathbf{Ay} \|_1$. Therefore,

$$f(\mathbf{A}) = \sum_{i=1}^{m} \left| \sum_{j=1}^{n} a_{ij} y_j \right| \le \sum_{i=1}^{m} \sum_{j=1}^{n} | a_{ij} | \cdot | y_j | =$$

$$= \sum_{j=1}^{n} \left(|y_j| \sum_{i=1}^{m} | a_{ij} | \right) \le \sum_{j=1}^{n} | y_j | \left(\max_{1 \le h \le n} \sum_{i=1}^{m} | a_{ih} | \right) =$$

$$= \| \mathbf{y} \|_1 \left(\max_{1 \le h \le n} \sum_{i=1}^{m} | a_{ih} | \right) = \max_{1 \le h \le n} \sum_{i=1}^{m} | a_{ih} | ,$$

which shows that all matrix norms subordinate to the minkowski 1-norm are bounded above by

$$\max_{1 \le h \le n} \sum_{i=1}^{m} | a_{ih} | .$$

If

$$\sum_{i=1}^{m} | a_{ik} | = \max_{1 \le h \le n} \sum_{i=1}^{m} | a_{ih} |,$$

then this bound is actually attained by $\|\mathbf{A}\, \mathbf{i}_{.,k}\|_1$, and since $\|\mathbf{i}_{.,k}\|_1 = 1$, we have

$$\max_{\|\mathbf{x}\|_1 = 1} \| \mathbf{Ax} \|_1 = \max_{1 \le h \le n} \sum_{i=1}^{m} | a_{ih} | .$$

qed

Theorem 2.14 A matrix norm subordinate to the minkowski ∞-norm is the maximum sum of the moduli of the entries in a row. For $\mathbf{A}_{m,n}$: $\| \mathbf{A} \|_\infty = \max_{1 \le i \le m} \sum_{j=1}^{n} | a_{ij} |$.

■

Proof: If f is a matrix norm subordinate to the minkowski ∞-norm, then there is a $\mathbf{y}$ such that $\| \mathbf{y} \|_\infty = 1$ and $f(\mathbf{A}) = \| \mathbf{Ay} \|_\infty$.
Therefore,

$$f(\mathbf{A}) = \max_{1 \le i \le m} \left| \sum_{j=1}^{n} a_{ij} y_j \right| \le \max_{1 \le i \le m} \left(\sum_{j=1}^{n} | a_{ij} | \; | y_j | \right) \le$$

$$\le \max_{1 \le h \le n} | y_h | \cdot \max_{1 \le i \le m} \sum_{j=1}^{n} | a_{ij} | =$$

$$= \| \mathbf{y} \|_\infty \max_{1 \le i \le m} \sum_{j=1}^{n} | a_{ij} | = \max_{1 \le i \le m} \sum_{j=1}^{n} | a_{ij} | ,$$

which shows that all matrix norms subordinate to the ∞-norm are bounded above by $\max_{1 \le i \le m} \sum_{j=1}^{n} | a_{ij} |$. If $\sum_{j=1}^{n} | a_{kj} | = \max_{1 \le i \le m} \sum_{j=1}^{n} | a_{ij} |$, then, assuming $\mathbf{A} \neq \mathbf{O}$, this bound is actually attained by $\| \mathbf{Az} \|_\infty$ where $\mathbf{z}$ is $\mathbf{a}_{k,\cdot}$ with the nonzero entries replaced by $a_{kj} / | a_{kj} |$. Since $\| \mathbf{z} \|_\infty = 1$, we have

$$\max_{\|\mathbf{x}\|_\infty = 1} \| \mathbf{Ax} \|_\infty = \max_{1 \le i \le m} \sum_{j=1}^{n} | a_{ij} | .$$

(Of course, if $\mathbf{A} = \mathbf{O}$ then $\| \mathbf{A} \| = 0$ for any matrix norm.)

qed

In our discussion of matrix norms, in particular the part about stretching capability, the reader may have been reminded of the definition of an eigenvalue. In the definition of eigenvalues, however, we were interested in the stretching of the vector without changing the direction of the vector. This additional constraint has consequences for the relation between a matrix' norm and its eigenvalues.

Theorem 2.15 No eigenvalue of $\mathbf{A}$ can be greater than $\| \mathbf{A} \|$:

$$\forall_{\lambda \in \sigma(A)} [\lambda \le \| \mathbf{A} \|]$$

■

Proof:

$$| \lambda | \, \| \mathbf{x} \| = \| \lambda \mathbf{x} \| = \| \mathbf{Ax} \| \le \| \mathbf{A} \| \, \| \mathbf{x} \|$$

qed

Before we go into the convergence properties of matrices we need one more result that may answer the following question: with so many different norms at our disposal which one should be used in establishing these properties? The answer is simply: whichever is the most convenient, because all norms turn out to be equivalent in the following sense.

Theorem 2.16 For each pair of seminorms f and g, that satisfy the additional requirement that the seminorm is zero fif all arguments are zero, there exist two positive real numbers m and M such that

$$\forall_{x\in\mathbb{C}^n}\ [m\, g(x) \le f(x) \le M\, g(x)]\,.$$

■

Proof: Let S be a unit sphere in $\mathbb{C}^n$: $S = \{x \mid \max_{1\le i\le n} |x_i| = 1\}$ i.e. a closed and bounded set of points. Since f is, like any seminorm, a continuous function (theorem 2.11), there must be a $y \in S$ and a $z \in S$ such that

$$\forall_{x\in S}\ [0 < f(y) \le f(x) \le f(z) < \infty]\,.$$

Of course, if $s(x)$ denotes $\max\limits_{1\le i\le n} |x_i|$, then

$$\forall_{x\in\mathbb{C}^n}\left[\frac{x}{s(x)} \in S\right],$$

and so

$$f(y) \le f\left(\frac{x}{s(x)}\right) \le f(z) \quad \text{or} \quad f(y)s(x) \le f(x) \le f(z)s(x).$$

Likewise for g there is $y', \in S$ and a $z' \in S$ such that

$$g(y')s(x) \le g(x) \le g(z')s(x), \quad \text{and so} \quad \frac{g(x)}{g(y')} \ge s(x) \ge \frac{g(x)}{g(z')}.$$

Consequently, $m = \frac{f(y)}{g(z')}$ and $M = \frac{f(z)}{g(y')}$ satisfy the conditions of the theorem.

qed

This concludes our small introduction into matrix norms. As mentioned before, the motivation for this introduction is to have that concept available when answering questions about the convergence of matrix expressions, an important topic in some of the chapters of this book. To avoid long detours when answering these questions in the proper context we include in this section some results concerning limits of matrices.

Definition 2.7 $\mathbf{A}(k)$ be some matrix whose entries depend on the non-negative integer k, then $\lim_{k \to \infty} \mathbf{A}(k) = \mathbf{A}$ if for every entry

$$\lim_{k \to \infty} a_{ij}(k) = a_{ij}.$$

■

Theorem 2.17

$$\lim_{k \to \infty} \mathbf{A}(k) = \mathbf{A} \iff \lim_{k \to \infty} \| \mathbf{A}(k) - \mathbf{A} \| = 0$$

■

Proof: If $\lim_{k \to \infty} \mathbf{A}(k) = \mathbf{A}$, then $\lim_{k \to \infty} (\mathbf{A}(k) - \mathbf{A}) = \mathbf{O}$. Thus, we have to prove that $\lim_{k \to \infty} \mathbf{B}(k) = \mathbf{O}$ fif $\lim_{k \to \infty} \| \mathbf{B}(k) \| = \mathbf{O}$.

Since any matrix norm is a seminorm, and therefore continuous, and since $\|\mathbf{O}\| = 0$, we immediately have that $\lim_{k \to \infty} \mathbf{B}(k) = \mathbf{O}$ implies that $\lim_{k \to \infty} \|\mathbf{B}(k)\| = \mathbf{O}$.

Conversely, if $\lim_{k \to \infty} \|\mathbf{B}(k)\| = \mathbf{O}$ for some matrix norm, then by theorem 2.16 we have $0 = \lim_{k \to \infty} M \|\mathbf{B}(k)\| \geq \lim_{k \to \infty} \|\mathbf{B}(k)\|$ which implies $\lim_{k \to \infty} \mathbf{B}(k) = \mathbf{O}$.

qed

Only square matrices are of interest to us, and therefore we often tacitly assume that the matrix is square.

Theorem 2.18 If some norm of $\mathbf{A}$ is strictly smaller than 1, then

$$\lim_{k \to \infty} \mathbf{A}^k = \mathbf{O}.$$

■

Proof: Obviously, $\|\mathbf{A}\| \leq \|\mathbf{A}\|$. Now, suppose $\|\mathbf{A}^{r-1}\| \leq \|\mathbf{A}\|^{r-1}$. Then, $\|\mathbf{A}^r\| = \|\mathbf{A}\mathbf{A}^{r-1}\| \leq \|\mathbf{A}\| \, \|\mathbf{A}^{r-1}\| \leq \|\mathbf{A}\| \, \|\mathbf{A}\|^{r-1} = \|\mathbf{A}\|^r$. Therefore, for $\|\mathbf{A}\| < 1$ we have
$0 \leq \lim_{k \to \infty} \|\mathbf{A}^k - \mathbf{O}\| = \lim_{k \to \infty} \|\mathbf{A}^k\| = 0$, and by theorem 2.17 $\lim_{k \to \infty} \|\mathbf{A}^k - \mathbf{O}\| = 0$ implies $\lim_{k \to \infty} \mathbf{A}^k = \mathbf{O}$.

qed

Theorem 2.19 $\forall_{r\in\mathbb{N}_0}\left[\lim_{k\to\infty} k^r\mathbf{A}^k = \mathbf{O}\right] \Longleftrightarrow \forall_{\lambda\in\bar{\sigma}(\mathbf{A})}\,[\,|\,\lambda\,| < 1\,]$. ∎

Proof: Suppose $\forall_{r\in\mathbb{N}_0}\left[\lim_{k\to\infty} k^r\mathbf{A}^k = 0\right]$. Then, by taking $r = 0$, we have $\lim_{k\to\infty}\mathbf{A}^k = \mathbf{O}$, and by theorem 2.17 $\lim_{k\to\infty}\|\mathbf{A}^k\| = 0$. Since no eigenvalue can be greater than any norm of its matrix (theorem 2.15), we have $0 = \lim_{k\to\infty}\|\mathbf{A}^k\| \geq \lim_{k\to\infty}|\,\lambda^k\,| = \lim_{k\to\infty}|\,\lambda|^k$ for all $\lambda\in\bar{\sigma}(\mathbf{A})$. Therefore, $\forall_{\lambda\in\bar{\sigma}(\mathbf{A})}[\,|\,\lambda\,| < 1]$.

Now, suppose all eigenvalues of $\mathbf{A}$ are strictly smaller than 1 in modulus. If $\mathbf{B} = \mathbf{S}\mathbf{A}\mathbf{S}^{-1}$ is in pseudo-diagonal normal form, then so is $\mathbf{S}\mathbf{A}^k\mathbf{S}^{-1}$, and $\mathbf{B}^k = \mathbf{S}\mathbf{A}^k\mathbf{S}^{-1}$, with the same block partitioning, and each block in $\mathbf{B}^k$ is the kth power of the corresponding block in $\mathbf{B}$. So, we can concentrate on the kth power of lower triangular matrices with all diagonal entries equal to the same eigenvalue λ of $\mathbf{A}$. If this eigenvalue is zero, then $\mathbf{B}_\lambda^k = \mathbf{O}$ for $k \geq m(\lambda)$ by theorem 2.1, and there is nothing left to prove. So we assume $\lambda \neq 0$:

$$\mathbf{B}_\lambda = \lambda\left(\mathbf{I}_{m(\lambda),\,m(\lambda)} + \mathbf{C}_{m(\lambda),\,m(\lambda)}\right)$$

where $\mathbf{C}$ is lower triangular with zeros on the diagonal. By theorem 2.1 again, $\mathbf{C}^k = \mathbf{O}$ for $k \geq m(\lambda)$. Therefore,

$$k^r\mathbf{B}_\lambda^k = \lambda^k k^r(\mathbf{I}+\mathbf{C})^k = \lambda^k k^r \sum_{h=0}^{k}\binom{k}{h}\mathbf{C}^h = \lambda^k k^r \sum_{h=0}^{m(\lambda)}\binom{k}{h}\mathbf{C}^h.$$

and all depends on whether $\lim_{k\to\infty}|\lambda|^k k^r\binom{k}{h} = 0$ for all $|\,\lambda\,| < 1$ and $h \leq m(\lambda)$. To prove this we use the following version of Stirling's formula

$$x! = \sqrt{2\pi}\,x^{x+\frac{1}{2}}\exp\left(-x + \frac{\vartheta}{12x}\right) \text{ with } 0 < \vartheta < 1.$$

Substituting this gives

$$\lim_{k\to\infty}|\,\lambda\,|^k k^r\binom{k}{h} =$$

$$= \tfrac{e^h}{h!}\lim_{k\to\infty}\left(|\,\lambda\,|^k k^{r+h}\right)\lim_{k\to\infty}\exp\left(\frac{\vartheta_1}{12k} - \frac{\vartheta_2}{12(k-h)}\right)\lim_{k\to\infty}\left(\tfrac{k}{k-h}\right)^{k-h+\frac{1}{2}}.$$

Since the limits on the right hand side are 0, 1 and 1, respectively, the limit on the left hand side must be 0.

qed

We conclude this section with a theorem that states that there is for every eigenvalue a diagonal element close to it if the off-diagonal elements are small enough in modulus.

Theorem 2.20 (Gershgorin)

If λ is an eigenvalue of $\mathbf{A}_{n,n}$, then

$$\exists_{1\leq j\leq n}\left[\,|\lambda - a_{jj}| \leq -|a_{jj}| + \sum_{i=1}^{n} |a_{ij}|\,\right]$$

■

Proof:

$$\exists_{\mathbf{x}\in\mathbb{C}^n}\,[\mathbf{A}\mathbf{x} = \lambda\mathbf{x} \wedge \|\mathbf{x}\|_\infty = 1]$$

Suppose $|x_r| = 1$, then $\sum_{j=1}^{n} a_{rj}x_j = \lambda x_r = \lambda$. Hence

$$|\lambda - a_{rr}| \leq \sum_{j\neq r} |a_{rj}x_j| \leq \sum_{j\neq r} |a_{rj}|\,|x_j| \leq \sum_{j\neq r} |a_{rj}|$$

qed

2.4 Quadratic forms

Definition 2.8 If $\mathbf{A}_{n,n}$ is a real symmetric matrix, then $\mathbf{x}_n^T\mathbf{A}_{n,n}\mathbf{x}_n$ is called the *quadratic form associated with* $\mathbf{A}$.

■

Using any nonsingular matrix $\mathbf{B}_{n,n}$ for the transformation $\mathbf{x}_n = \mathbf{B}_{n,n}\mathbf{y}_n$ yields another quadratic form:

$$\mathbf{x}^T\mathbf{A}\mathbf{x} = (\mathbf{B}\mathbf{y})^T\mathbf{A}(\mathbf{B}\mathbf{y}) = \mathbf{y}^T\mathbf{B}^T\mathbf{A}\mathbf{B}\mathbf{y}$$

The matrix $\mathbf{B}^T\mathbf{A}\mathbf{B}$ is clearly symmetric, since $(\mathbf{B}^T\mathbf{A}\mathbf{B})^T = \mathbf{B}^T\mathbf{A}^T\mathbf{B} = \mathbf{B}^T\mathbf{A}\mathbf{B}$. The two matrices, $\mathbf{A}$ and $\mathbf{B}^T\mathbf{A}\mathbf{B}$, are called *congruent* and their associated quadratic forms are called *equivalent*. Some of the equivalent quadratic forms may have a simple and compact representation. The following theorem states that there is simple equivalent quadratic form for any given quadratic form.

Theorem 2.21 If a quadratic form x^TAx is not identically zero, then there exists a nonsingular matrix B such that the transformation $x = By$ yields a quadratic form that is just a sum of squares:

$$\forall_{A_{n,n} \neq O_{n,n}} \exists_{\det(B_{n,n}) \neq 0} \exists_{1 < r \leq n} \left[y^T B^T A B y = \sum_{i=1}^{r} d_i y_i^2 \wedge \forall_{1 \leq i \leq r} [|d_i| = 1] \right]$$

The integer r, called the rank of the quadratic form, and also of the matrix A, denoted by rank(A), and the integer $\sum_{i=1}^{r} d_i$, called the *signature* of the quadratic form, and also of the matrix A, denoted by signature(A), are invariant over the set of equivalent forms and congruent matrices.

■

Proof: The proof will be by induction. When $n = 1$ the required transformation will be $y_1 = (\sqrt{|a_{11}|})x_1$.

Now suppose that the theorem is true for quadratic forms in $n-1$ or fewer variables. There are a few cases to distinguish.

If all diagonal elements of A are zero, then we select an arbitrary nonzero element, say a_{hk}, $h < k$, and we apply the nonsingular transformation $x = Cz$ where C is equal to the identity matrix, except for the entry c_{kh}, which is 1. As a result of that transformation the terms $a_{hk}x_hx_k$ and $a_{kh}x_kx_h$ are replaced by $a_{hk}z_h^2 + a_{hk}z_hz_k$. The entry $(C^TAC)_{hh}$ will be equal to $2a_{hk}$, and therefore nonzero. Thus, the case of a zero diagonal in A is transformed into the next case.

There is a lowest index k, with $a_{kk} \neq 0$. Suppose, however, that $k \neq 1$. The nonsingular transformation $x = Cz$ where C is equal to the identity matrix, except for the entries c_{11}, c_{1k}, c_{k1}, and c_{kk}, which are 0, 1, 1, and 0, respectively, transforms this case into the next one.

Assume $a_{11} \neq 0$, then

$$
\begin{aligned}
\mathbf{x}^T\mathbf{A}\mathbf{x} &= \sum_{i=1}^{n}\sum_{j=1}^{n} a_i x_{ij} a_j = a_{11}x_1^2 + 2\sum_{j=2}^{n} a_{1j}x_1x_j + \sum_{i=2}^{n}\sum_{j=2}^{n} a_{ij}x_ix_j = \\
&= a_{11}\left(x_1^2 + 2x_1\sum_{j=2}^{n}\frac{a_{1j}}{a_{11}}x_j\right) + \sum_{i=2}^{n}\sum_{j=2}^{n} a_{ij}x_ix_j = \\
&= a_{11}\left(x_1^2 + 2x_1\sum_{j=2}^{n}\frac{a_{1j}}{a_{11}}x_j + \left(\sum_{j=2}^{n}\frac{a_{1j}}{a_{11}}x_j\right)^2\right) + \\
&\qquad + \sum_{i=2}^{n}\sum_{j=2}^{n} a_{ij}x_ix_j - a_{11}\left(\sum_{j=2}^{n}\frac{a_{1j}}{a_{11}}x_j\right)^2 = \\
&= a_{11}\left(x_1 + \sum_{j=2}^{n}\frac{a_{1j}}{a_{11}}x_j\right)^2 + \sum_{i=2}^{n}\sum_{j=2}^{n} a_{ij}x_ix_j - a_{11}\left(\sum_{j=2}^{n}\frac{a_{1j}}{a_{11}}x_j\right)^2 .
\end{aligned}
$$

The part after the last plus sign is a quadratic form in the variables $x_2 \ldots x_n$. By our induction hypothesis, there exists a matrix $\mathbf{C}'_{n-1,\,n-1}$ reducing that quadratic form to the desired sum of squares. Therefore, the transformation required in the theorem is $\mathbf{y} = \mathbf{CDx}$ where $\mathbf{D}$ is equal to the identity matrix, except for the first row for which $\mathbf{D}_{1,\cdot} = \frac{A_{1,\cdot}}{a_{11}}$, and

$$
\mathbf{C} = \left[\begin{array}{c|ccc}
\sqrt{|a_{11}|} & 0 & \cdots & 0 \\
\hline
0 & & & \\
\vdots & & \mathbf{C}' & \\
0 & & &
\end{array}\right] .
$$

The remainder of the theorem is obvious after observing that the number of positive coefficients and the number of negative coefficients in any sums-of-squares representation is invariant under nonsingular linear transformations. A simple reasoning, reducing the negation to absurdity, shows that. These numbers uniquely determine the rank and the signature.

qed

Further, since it is obvious that when $\mathbf{A}$ is nonsingular, so is $\mathbf{B}^T\mathbf{A}\mathbf{B}$, and since the latter is a diagonal matrix, it follows that

Corollary 2.1 The rank of $\mathbf{x}^T\mathbf{A}\mathbf{x}$ is n fif $\mathbf{A}$ is nonsingular.

■

Definition 2.9 A matrix, and also its associated quadratic form, is said to be *positive semi-definite* when its rank is equal to its signature. They are called *positive definite* when order, rank and signature are equal.

■

It easily follows that

Theorem 2.22

$$\begin{aligned} \mathrm{rank}(\mathbf{A}_{n,n}) &= \mathrm{signature}(\mathbf{A}_{n,n}) \Longleftrightarrow \forall_{x \in \mathbb{R}^n}[\mathbf{x}^T\mathbf{A}\mathbf{x} \geq 0] \\ n = \mathrm{rank}(\mathbf{A}_{n,n}) &= \mathrm{signature}(\mathbf{A}_{n,n}) \Longleftrightarrow \forall_{x \in \mathbb{R}^n \setminus \{o_n\}}[\mathbf{x}^T\mathbf{A}\mathbf{x} > 0] \end{aligned}$$

■

If a real symmetric matrix $\mathbf{A}$ is positive definite there is a matrix $\mathbf{B}$ such that $\mathbf{B}^T\mathbf{A}\mathbf{B} = \mathbf{I}$.

Corollary 2.2 A real symmetric matrix is positive definite fif there exists a non-singular matrix $\mathbf{B}$ such that $\mathbf{A} = \mathbf{B}^T\mathbf{B}$

■

Theorem 2.23 If $\mathbf{A}_{n,n}$ is positive definite real symmetric matrix, $\mathbf{x}_n$ a real vector, and $\mathbf{c}_n$ a complex vector independent from $\mathbf{x}$, then

$$\int_{\mathbb{R}^n} \exp\left(-\tfrac{1}{2}(\mathbf{x}-\mathbf{c})^T\mathbf{A}^{-1}(\mathbf{x}-\mathbf{c})\right) dx_1 \ldots dx_n = \sqrt{2^n\pi^n \det(\mathbf{A})}.$$

■

Proof: Since $\mathbf{A}$ is positive definite, there exists a nonsingular matrix $\mathbf{R}_{n,n}$, such that $\mathbf{A} = \mathbf{R}^T\mathbf{R}$. We substitute $\mathbf{y}_n$ for $\mathbf{x} - \mathbf{c}$ and then change variables by the transformation $\mathbf{z}_n = (\mathbf{R}^T)^{-1}\mathbf{y}$.

$$\int_{\mathbb{R}^n} \exp\left(-\tfrac{1}{2}(\mathbf{x}-\mathbf{c})^T\mathbf{A}^{-1}(\mathbf{x}-\mathbf{c})\right) dx_1 \dots dx_n =$$

$$= \int_{\mathbb{R}^n} \exp\left(-\tfrac{1}{2}\mathbf{y}^T\mathbf{A}^{-1}\mathbf{y}\right) dy_1 \dots dy_n =$$

$$= \int_{\mathbb{R}^n} \exp\left(-\tfrac{1}{2}\mathbf{z}^T\mathbf{A}^{-1}\mathbf{z}\right) \det(\mathbf{R}) dz_1 \dots dz_n =$$

$$= \sqrt{\det(\mathbf{A})} \left(\int_{-\infty)}^{\infty} \exp\left(-\tfrac{1}{2}\mathbf{z}^T\mathbf{z}\right) dz\right)^n .$$

Since

$$\left(\int_{-\infty}^{\infty} \exp\left(-\tfrac{1}{2}z^2\right) dz\right)^2 =$$

$$= \int_{-\infty}^{\infty}\int_{-\infty}^{\infty} \exp\left(-\tfrac{1}{2}\left(z_1^2 + z_2^2\right)\right) dz_1 dz_2 =$$

$$= \int_0^{2\pi}\int_0^{\infty} r \exp\left(-\tfrac{1}{2}r^2\right) dr d\theta = 2\pi,$$

the n-fold integral above must have the value $\sqrt{\det(\mathbf{A})}\,\left(\sqrt{2\pi}\right)^n$.

qed

2.5 Discussion

The facts collected in section 2.1 can be found in any good introduction into matrix theory. Pseudo-diagonal normal forms are sometimes included, but seldom part of a core-curriculum. The solution of the matrix equation, crucial to the derivation of these normal forms in section 2.2, is from [129] Stephanos' rule was published in 1900, and Neudecker's identity can be found in [115]. The material on norms is more or less standard. More extensive treatments are sometimes found in books on numerical methods, in particular specialized on numerical methods for linear algebra. The proof of theorem 2.21 is called the lagrange reduction method. The last part is known as the inertia law of Sylvester.

3 CHAINS

The inner loop of an annealing algorithm is a markov process, and since the value of t is not changed within this loop the process is also homogeneous. A markov process is called a markov chain when a sequence of events rather than a process defined in continuous time is described. Finally, the fact that the number of states is finite makes the loop a *finite homogeneous markov chain.* Matrices provide a convenient formalism for discussing these chains, and we therefore describe the inner loop of the annealing algorithm as a sequence of matrix multiplications. Many results are derived using the matrix formulation of the chain, but the central result of this chapter is that under certain mild conditions the frequencies of the states in the loop will tend to a stationary density function, called the *equilibrium density* of the chain. The equilibrium density depends on t.

It is an important result, because we want the state densities close to the equilibrium density while running the annealing algorithm. That is what is meant by the phrase *maintaining quasi-equilibrium.* Knowledge of the equilibrium density helps us to find ways of controlling the process in such a way that quasi-equilibrium is maintained. The result in this chapter is still quite general, but as more becomes known about the selection and acceptance function more specific properties of the equilibrium density can be derived from it. These more specific results will be topics in chapters 5 and 8. In this chapter we only study a slight specification that is very common in many problem formulations for applying the annealing algorithm , namely *reversability.* Reversible chains have surprising properties. A number of these properties are derived at the end of this chapter.

3.1 Terminology

An annealing algorithm works on a *state space*, which is a set with a relation. The elements of the set are called *states*. Each state represents a configuration. Since each state must be encoded such that it can be distinguished from any other state by a computer, there is no loss of generality when we require this set to be finite. We denote the set by $\mathcal{S}$ and its cardinality by s. A *score function*, $\varepsilon : \mathcal{S} \to \mathbb{R}_+$, assigns a positive real number to each state. This number is interpreted as a quality indicator, in the sense that the lower this number the better is the configuration that is encoded in that state. By defining a set μ of neighbor relations over $\mathcal{S}$ (i.e. $\mu \subseteq \mathcal{S} \times \mathcal{S}$) a *structure*, sometimes called a *topology*, is endowed to the state set $\mathcal{S}$. The elements of μ are called *moves*, and the states in $(s, s') \in \mu$ are said to be connected via a single move, or simply *adjacent*. Similarly, $(s, s') \in \mu^k$ are said to be connected via a set of k moves. Since we want any state to be connected to any other state by a finite number of moves, we require the transitive closure of μ to be the universal relation over $\mathcal{S}$:

$$\bigcup_{k=1}^{\infty} \mu^k = \mathcal{S} \times \mathcal{S} \tag{3.1}$$

The *diameter* $\phi(\mathcal{S}, \mu)$ of a space $\mathcal{S}$ with move set μ satisfying (3.1), is defined as the smallest integer such that

$$\forall_{(s,s') \in \mathcal{S} \times \mathcal{S}} \exists_{k \leq \phi(\mathcal{S},\mu)} \left[(s, s') \in \mu^k \right].$$

This implies that

$$\bigcup_{k=1}^{\phi(\mathcal{S},\mu)} \mu^k = \mathcal{S} \times \mathcal{S}.$$

The set of states connected with a given state s, is denoted by $s\mu$, and its cardinality is called the degree of s. We do not exclude move sets in which a state is adjacent to itself. To treat such moves differently, we introduce for convenience the notation ι for the *identity relation*: $\iota = \{(s, s) \mid s \in \mathcal{S}\}$.

Two numbers are associated with each pair of states. One of them is called the *selection probability* $\beta : \mathcal{S} \times \mathcal{S} \rightarrow [0,1] \subset \mathbb{R}$. Only pairs in μ should be selected, and therefore

$$\forall_{(s,\, s') \notin \mu} [\beta(s,\, s') = 0] \tag{3.2}$$

$$\forall_{(s,\, s') \in \mu} [\beta(s,\, s') \neq 0] \tag{3.3}$$

$$\forall_{s \in \mathcal{S}} \left[\sum_{s' \in \mathcal{S}\mu} \beta(s,\, s') = 1 \right] \tag{3.4}$$

Note that the selection probability is never zero for a pair of states connected by a single move. Another function, called the *acceptance function*, assigns a positive probability measure to a pair of scores, and a positive real number, the *control parameter*. Therefore,

$$\alpha : \mathbb{R}_+^{\,3} \rightarrow (0,1] \subset \mathbb{R}.$$

The state set $\mathcal{S}$, the relation μ, and the functions α and β completely determine what we will call a *chain*. When defined in the form of a (pseudo) computer program it is nothing else than a never-ending loop such as the one given in the procedure chain.

```
PROCEDURE chain(t);
BEGIN cstate := generate;
      cscore := epsilon(cstate) ;
      WHILE true DO
      BEGIN nstate := select(cstate, beta(cstate));
            nscore := epsilon(nstate) ;
            IF  random < alpha(cscore, nscore, t)  THEN
            BEGIN cstate := nstate ;
                  cscore := nscore ;
            END;
      END;
END;
```

The conditional probability

$$\mathsf{P}(\texttt{nstate} = s' \mid \texttt{cstate} = s)$$

is called the *transition probability*, and denoted by $\tau(s, s', \mathsf{t})$. From the definition of chains we immediately obtain

$$\tau(s, s', \mathsf{t}) = \alpha(\varepsilon(s), \varepsilon(s'), \mathsf{t})\, \beta(s, s')$$

when $s \neq s'$, and

$$\tau(s, s, \mathsf{t}) = 1 - \sum_{s \neq s'} \alpha(\varepsilon(s), \varepsilon(s'), \mathsf{t})\, \beta(s, s').$$

This probability only depends on the two states involved and the external parameter t. This property is called the *markov property*, and processes possessing this property are called *markov processes*. Our chain is therefore an example of a markov process.

Each time the body of the loop is completed we say that the chain has completed a step. Each step has its own `cstate`. The conditional probability that $s' =$`cstate` given that n steps earlier $s =$ `cstate` is denoted by $\tau_{\mathsf{n}}(s, s', \mathsf{t})$, and implied by τ, because

$$\tau_{\mathsf{n}}(s, s', \mathsf{t}) = \sum_{s'' \in \mathcal{S}} \tau_{\mathsf{n}-1}(s, s'', \mathsf{t})\, \tau_1(s'', s', \mathsf{t})$$

where $\tau_1 = \tau$ and $\tau_0(s, s', \mathsf{t}) = 0$ except for $s = s'$ in which case $\tau_0(s, s, \mathsf{t}) = 1$.

3.2 Linear arrangement, an example

Before deriving some properties of chains, let us introduce an example. Suppose we have a number of equally sized, say square objects, often called *cells* or *modules*, and a list of interconnections, often called *nets*. For each net a number of objects are mentioned in this list. These objects have to be interconnected by that net. All the objects have to be placed next to each other with their centers on a given straight line segment of minimum length. A result of such a placement is called a *linear arrangement*. For each net we calculate the length of the minimum interval on that line, that contains all the centers of the modules to be interconnected by that net. We will call that the length of that net. We

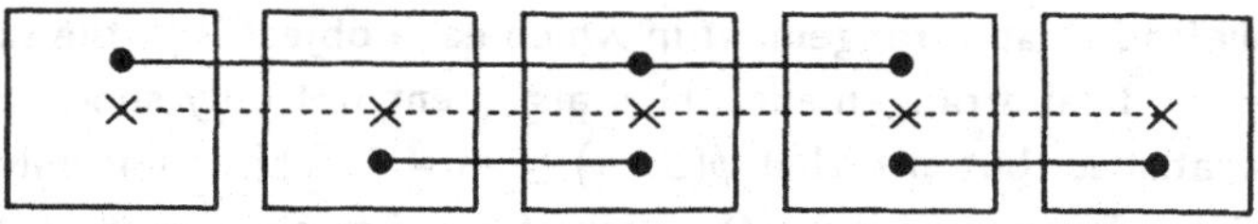

Figure 3.1: The linear arrangement problem

would like to find an arrangement where the sum of all the net lengths is small, preferably minimum.

The combination of a set of objects with a net list is called an *incidence structure.* There is the set $\mathcal{M}$ of m modules and the set $\mathcal{N}$ of n nets, and the mechanism to relate modules with nets, for example a relation over $\rho \subseteq \mathcal{N} \times \mathcal{M}$. The elements of ρ are often called *pins.* There is an obvious one-to-one relationship between the permutations of the modules in $\mathcal{M}$ and the arrangements described in the previous paragraph. We therefore choose the set of all permutations of the m objects in $\mathcal{M}$ as the set of states of a chain. The score function ε assigns to each permutation the sum of the net lengths in the associated arrangement. If we view each state as a bijective function $s : \mathcal{M} \to \{1, \ldots, m\}$, we can write the score function as

$$\varepsilon(s) = \sum_{n \in \mathcal{N}} \left(\max_{m \in n\rho} s(m) - \min_{m' \in n\rho} s(m') \right).$$

The representation of each arrangement and the definition of the objective function are quite obvious, considering the given problem. This is not the case with the move set. Selecting a move set, as we will see later, is an important step in constructing a state space. Here we will choose the identity and the transpositions as moves. In other words, two states are adjacent if one can be obtained from the other by interchanging two objects in the associated arrangement or leaving everything unchanged. This move set satisfies the requirement that every state has to be connected with any other state by a finite set of moves (3.1). To verify this, consider two arbitrary arrangements or equivalently their associated permutations, an initial one and a final one. Objects that are in the same position in the two arrangements can be left alone. For an object that is in a position different from its position in the final arrangement one needs a single transposition to put that object in the position that it occupies in the final arrangement. From then on that object is left in that position. Note that there cannot be only one object in a non-final position. Repeating the procedure for each ensuing arrangement must ultimately lead to an arrangement in which each object is in the same position as given by the final arrangement. This argument not only shows that requirement (3.1) is satisfied, but also that $\phi(\mathcal{S}, \mu) \leq m - 1$. This is not only an upper bound, but also a lower bound for the diameter . For each permutation can be uniquely decomposed in disjoint cycles. If a transposition is applied to a permutation, the number of cycles in its decomposition changes by 1. Since there is a permutation with 1 cycle and a permutation with m cycles, it takes at least $m - 1$ transpositions to carry the former over into the other. Therefore, $\phi(\mathcal{S}, \mu) = m - 1$. We can also determine the degree of a state in this space. For a given state we can choose $\frac{1}{2}m(m - 1)$ different object pairs, each one leading to a different state. If we include the possibility of staying in the same state we have $\frac{1}{2}m^2 - \frac{1}{2}m + 1$ distinct states adjacent to the given state. Consequently, the degree of all states in $(\mathcal{S}, \mu)$ is $\frac{1}{2}m^2 - \frac{1}{2}m + 1$.

We still have to specify the selection function β. Since there is for now no obvious reason to make the numbers assigned to moves involving two distinct states different, we choose

$$\forall_{(s,\, s') \in \mu \setminus \iota} \left[\beta(s,\, s') = 2\frac{1 - a}{m(m - 1)} \right]$$

where a is the number assigned to the identity move by β. It is convenient for implementations to have $a = \frac{1}{m}$, but regardless of the value of a (as long as $0 < a < 1$) the requirements (3.2), (3.3), (3.4) are satisfied.

3.3 The chain limit theorem

We can collect the transitions probabilities into a matrix $\mathbf{T}_{s,s}$, the so-called *transition matrix*. For this purpose we give all states a unique integer index. The entries of $\mathbf{T}$ are $t_{ij} = \tau(s_i, s_j, t)$. Some properties of this matrix can be immediately discovered:

$$\forall_{1 \leq i \leq s} \forall_{1 \leq j \leq s} \left[t_{ij} \geq 0 \right]$$

$$\forall_{1 \leq i \leq s} \left[\sum_{j=1}^{s} t_{ij} = 1 \right]$$

$$\forall_{1 \leq j \leq s} \exists_{1 \leq i \leq s} \left[t_{ij} \neq 0 \right].$$

Definition 3.1 A square matrix whose entries are non-negative, and whose row sums are equal to unity, is called *stochastic*. (Sometimes the column sums are, in addition, required to be not zero.)

■

$\mathbf{T}$ is clearly stochastic. The second property can also be written as $\mathbf{Tj}_s = \mathbf{j}_s$ so we have immediately that unity is an eigenvalue of $\mathbf{T}$ and $\mathbf{j}$ is its associated right eigenvector. Eigenvalues of $\mathbf{T}$ are in modulus not greater than 1. This follows immediately from theorem 2.15, because $\|\mathbf{T}\|_\infty = \sum_{j=1}^{s} t_{ij} = 1$. This is generally valid for stochastic matrices, and thus:

Theorem 3.1 A stochastic matrix $\mathbf{A}_{n,n}$ has an eigenvalue equal to unity, and the associated right eigenvector is $\mathbf{j}_n$. Further, no eigenvalue of $\mathbf{A}$ is in modulus greater than 1. In short:

$$\forall_{\lambda \in \bar{\sigma}(\mathbf{A})} [\,|\,\lambda\,| \leq 1], \quad 1 \in \bar{\sigma}(\mathbf{A}), \quad \mathbf{Aj} = \mathbf{j}$$

■

It is clear from the context from which we assembled $\mathbf{T}$ that the product of two stochastic matrices is again stochastic. Also the formal proof is easy:

Theorem 3.2 The product of two stochastic matrices is a stochastic matrix.

■

Proof: Suppose $\mathbf{A}_{n,n}$ and $\mathbf{B}_{n,n}$ are both stochastic matrices, and $\mathbf{C} = \mathbf{AB}$. Since each entry of $\mathbf{C}$ is the sum of products of nonnegative factors, each entry has to be nonnegative. Also the row sums of $\mathbf{C}$ equal unity, for

$$\sum_{j=1}^{n} c_{ij} = \sum_{j=1}^{n} \sum_{k=1}^{n} a_{ik} b_{kj} = \sum_{k=1}^{n} a_{ik} \sum_{j=1}^{n} b_{kj} = \sum_{k=1}^{n} a_{ik} = 1$$

qed

Of course, we can also collect the n-step transition probabilities τ_n into a matrix of order s, but considering the definition of τ_n it is easy to see that this matrix would be equal to the nth power of $\mathbf{T}$. We therefore do not introduce a separate notation for these matrices.

We have already noted that a transition matrix cannot have an eigenvalue whose modulus exceeds 1, but it may have several distinct eigenvalues whose modulus is equal to 1. An important class, however, is formed by the chains that do not have any eigenvalue whose modulus is equal to 1 while its value is not equal to 1. All the chains we will consider in this book, for example, belong to this class. Gersgorin's theorem 2.20 can be used to obtain quite easily a sufficient condition for the absence of modulus-1-eigenvalues other than 1 itself. Namely, if $w = \min_{1 \le i \le s} t_{ii}$ we have by that theorem

$$\forall_{\lambda \in \bar{\sigma}(T)} [\lambda \in \{z \mid z \in \mathbb{C} \wedge |z - w| \le 1 - w\}] \tag{3.5}$$

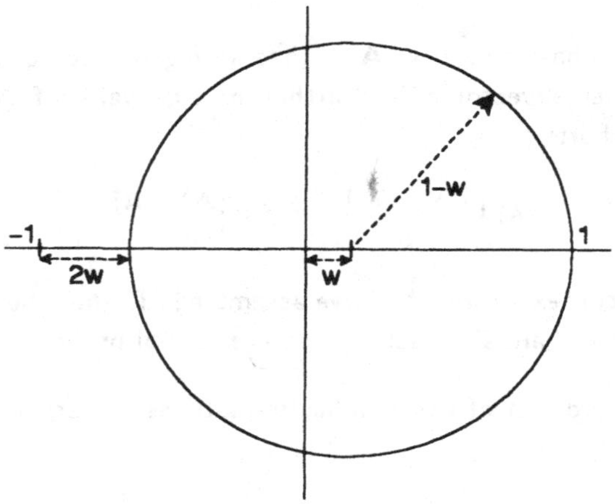

If $w > 0$ the only point with modulus 1 contained in the specified closed disk is 1. Transition matrices with no zero on the diagonal belong to chains with a reflexive move set (formally: $\iota \subseteq \mu$).

Definition 3.2 Chains that have a transition matrix with no zero on the diagonal are called *reflexive chains*.

■

With this definition we can formulate the result of the discussion preceding it as follows:

Lemma 3.1 If T is the transition matrix of a reflexive chain then

$$\forall_{\lambda \in \bar{\sigma}(A)} [|\lambda| = 1 \Rightarrow \lambda = 1]$$

■

Another property of transition matrices of reflexive chains is that by multiplying such a matrix repeatedly with itself all zero entries will disappear. This follows from the observation that if $(s_i, s_j) \in \mu^k$ the entry b_{ij} in $\mathbf{B} = \mathbf{T}^p$ will be nonzero for any $p \geq k$. This is obvious for $p = k$ because β does not assume the value zero over μ, and α is always positive. That it is true for all $p > k$ is a consequence from the fact that $t_{ii} \neq 0$ in the transition matrix of a reflexive chain. The requirement (3.1) guarantees that $(s_i, s_j) \in \mu^k$ for some $k \leq \phi(S, \mu)$. Therefore, $\mathbf{T}^p$ will not have a single zero entry for any $p \geq \phi(S, \mu)$. Let us summarize this also in a lemma.

Lemma 3.2 If $\mathbf{T}$ is the transition matrix of a reflexive chain and $\mathbf{B} = \mathbf{T}^p$ then $p \geq \phi(S, \mu) \Rightarrow \forall_{1\leq i\leq s} \forall_{1\leq j\leq s} \left[b_{ij} \neq 0\right]$.

■

Whereas the above results can be obtained easily from theorems that are themselves simple and easy to prove, the next one requires somewhat more effort, even though most of the work has already been done in section 2.2. What we want to establish is the fact that the eigenvalue $\lambda = 1$ of the transition matrix of a reflexive chain has multiplicity 1. Let us first prove the following lemmas.

Lemma 3.3 The block of eigenvalue 1 in a pseudo-diagonal normal form of the transition matrix of a reflexive chain has no non-zero off-diagonal entry.

■

Proof: Let $\mathbf{T}$ be the transition matrix of a reflexive chain, and let $\mathbf{S}$ be a non-singular matrix such that $\mathbf{A} = \mathbf{S}^{-1}\mathbf{T}\mathbf{S}$ is in pseudo-diagonal normal form (theorem 2.10). Further, let the ith row be the first row in $\mathbf{A}$ with a nonzero entry off the diagonal in the block of eigenvalue 1. Finally, suppose that that entry is in the jth column. Then $a_{ij} \neq 0$ and if $\mathbf{B} = \mathbf{A}^k$, then $b_{ij} = k\,a_{ij}$. This implies that

$$| k\,a_{ij} | \leq \|\mathbf{B}\|_\infty = \|\mathbf{A}^k\|_\infty =$$

$$= \|\mathbf{S}^{-1}\mathbf{T}^k\mathbf{S}\|_\infty \leq \|\mathbf{S}^{-1}\|_\infty \|\mathbf{T}^k\|_\infty \|\mathbf{S}\|_\infty \leq \|\mathbf{S}^{-1}\|_\infty \|\mathbf{T}\|_\infty^k \|\mathbf{S}\|_\infty =$$

$$= \|\mathbf{S}^{-1}\|_\infty \|\mathbf{S}\|_\infty$$

and since the last product is a constant independent from k, it cannot be greater or equal than $| k\,a_{ij} |$ for all k when $a_{ij} \neq 0$.

qed

Lemma 3.4 The right eigenvectors of the transition matrix of a reflexive chain associated with an eigenvalue equal to 1 are multiples of **j**.

■

Proof: Let $\mathbf{T}_{s,s}$ be the transition matrix of a reflexive chain, and let $\mathbf{v}_s$ be a right eigenvector associated with the eigenvalue 1. Further, let $\mathbf{B} = \mathbf{T}^k$ with $k \geq \phi(\mathcal{S}, \mu)$. This implies that

$$\mathbf{Tv} = \mathbf{v} = \mathbf{Bv}.$$

There must be a component of $\mathbf{v}$, say v_i, such that $|v_i| = \|\mathbf{v}\|_\infty$. So, we have

$$\|\mathbf{v}\|_\infty = |\sum_{j=1}^{s} b_{ij} v_j| \leq \sum_{j=1}^{s} b_{ij} |v_j| \leq \|\mathbf{v}\|_\infty \sum_{j=1}^{s} b_{ij} = \|\mathbf{v}\|_\infty.$$

Because of the equality of the members at the extreme ends, we can replace the inequality symbols by equality symbols, and conclude that $b_{ij} \neq 0 \Rightarrow v_i = v_j$. According to lemma 3.2, **B** does not have a zero entry, and therefore, $\mathbf{v} = \|\mathbf{v}\|_\infty \mathbf{j}_n$.

qed

We can summarize the preceding lemmas in the following theorem:

Theorem 3.3 The transition matrix of a reflexive chain has a simple eigenvalue equal to unity with eigenvector **j**. All other eigenvalues have modulus smaller than 1.

■

Proof: Suppose that **T** is the transition matrix of a reflexive chain, and the multiplicity of the eigenvalue 1 is greater than 1. Let **S** be a nonsingular matrix such that $\mathbf{A} = \mathbf{S}^{-1}\mathbf{TS}$ is in pseudo-diagonal normal form and $a_{11} = a_{22} = 1$ (theorem 2.10). Then, it easily follows, that $\mathbf{i}_{\cdot,1}$ and $\mathbf{i}_{\cdot,2}$ are two independent right eigenvectors of **T** associated with eigenvalue 1. However, this contradicts with lemma 3.4. Therefore, the multiplicity of eigenvalue 1 has to be 1. The remaining part of the theorem is in lemma 3.4 and theorem 3.1.

qed

The following theorem is the most important result of this section. Translated into the terminology of chains we will call it the *chain limit theorem*. Its matrix formulation is as follows:

Theorem 3.4 If $\mathbf{T}$ is the transition matrix of a reflexive chain, then the limit of $\mathbf{T}^k$ for $k \to \infty$ exists and

$$\mathbf{T}^\infty := \lim_{k \to \infty} \mathbf{T}^k = \mathbf{j}\mathbf{d}^T$$

where $\mathbf{d}_s$ is the unique standardized left eigenvector of $\mathbf{T}$ associated with the eigenvalue 1. All components of $\mathbf{d}$ are strictly positive.

■

Proof: Let $\mathbf{S}$ be a nonsingular matrix such that $\mathbf{A} = \mathbf{S}^{-1}\mathbf{T}\mathbf{S}$ is in pseudo-diagonal normal form and $a_{11} = 1$. Then

$$\lim_{k \to \infty} \mathbf{T}^k = \lim_{k \to \infty} \mathbf{S}\mathbf{A}^k\mathbf{S}^{-1} = \mathbf{S}\left(\lim_{k \to \infty} \mathbf{A}^k\right)\mathbf{S}^{-1} = \mathbf{S}\mathbf{B}\mathbf{S}^{-1}$$

where $\mathbf{B}$ is a matrix with $b_{11} = 1$ and all other entries equal to zero. This is a consequence of the fact that the block of eigenvalue 1 in $\mathbf{A}$ is the identity matrix of order 1, while all the other blocks are associated with an eigenvalue whose modulus is less than 1, and therefore converge to $\mathbf{O}$ as k goes to ∞ (theorem 2.19). So, the limit exists and with $\mathbf{R} = \mathbf{S}^{-1}$ we have $\mathbf{T}^\infty = \mathbf{S}_{.,1}\mathbf{R}_{1,.}$. Obviously, $\mathbf{T}\mathbf{T}^\infty = \mathbf{T}^\infty = \mathbf{T}^\infty\mathbf{T}$ which shows that every row of $\mathbf{T}^\infty$ must be a left eigenvector of $\mathbf{T}$ and every column must be a right eigenvector of $\mathbf{T}$, all associated with the eigenvalue 1. Since 1 is a simple eigenvalue, the columns have to be multiples of $\mathbf{j}$ (theorem 3.3), and the rows have to be multiples of each other. Moreover, since $\mathbf{T}^\infty$ is stochastic the rows must be equal to the transpose of the (unique) standardized left eigenvector of $\mathbf{T}$.

The positivity of the components of $\mathbf{d}$ follows from lemma 3.2.

qed

Regardless of which state is selected before initiation of the loop, any state will be the current state in some step, because of requirement (3.1) and the fact that neither α nor β can assume the value 0 over the set of moves. If the relation μ is required to be reflexive, i.e.

$$\forall_{s \in \mathcal{S}} [(s,\, s) \in \mu], \tag{3.6}$$

then, because of (3.3) and (3.4), $\tau(s,\, s) > 0$ for any state s. Chains with that property are, in accordance with definition 3.2 called *reflexive chains.* For these chains we have the important theorem 3.4 which can also be stated in the terminology of chains instead of transition matrices:

Theorem 3.5 For each reflexive chain there exists a positive density function $\delta : \mathcal{S} \times \mathrm{IR}_+ \to (0,1]$ defined by

$$\delta(s,\, \mathrm{t}) = \lim_{\mathrm{n} \to \infty} \tau_{\mathrm{n}}(s',\, s,\, \mathrm{t}) \tag{3.7}$$

(the limit does not depend on s'), and satisfying the relations

$$\delta(s,\, \mathrm{t}) = \sum_{s' \in \mathcal{S}} \delta(s',\, \mathrm{t})\tau(s',\, s,\, \mathrm{t}) \tag{3.8}$$

$$\sum_{s \in \mathcal{S}} \delta(s',\, \mathrm{t}) = 1. \tag{3.9}$$

■

Closely related to the equilibrium density δ is the *score density*

$$\hat{\delta}(\varepsilon,\, \mathrm{t}) = \sum_{\{s | \varepsilon(s) = \varepsilon\}} \delta(s,\, \mathrm{t}). \tag{3.10}$$

To appreciate the meaning of the theorem, consider a large enough number N of copies of the same reflexive chain started simultaneously in arbitrary initial states. After a sufficient number of steps the number of chains with current state s will be approximately $\mathrm{N}\,\delta(s,\, \mathrm{t})$. If the initial states are distributed according to the density function δ, then this density will be maintained at all steps. The function δ is therefore called the *equilibrium density* of the chain. The conditions of the statement also imply that a chain *in equilibrium* yields an ergodic process, in the sense that the relative state frequencies in infinitely many identical chains at a given step are the same as the relative state frequencies in a single such chain over infinitely many steps.

If we use the fact that $\sum_{s \in \mathcal{S}} \tau(s, s', \mathrm{t}) = 1$ we can rewrite the relations for δ as

$$\begin{aligned}\delta(s, \mathrm{t}) &= \sum_{s' \in \mathcal{S}} \delta(s', \mathrm{t})\tau(s', s, \mathrm{t}) \\ &= \sum_{s' \in \mathcal{S} \setminus \{s\}} \delta(s', \mathrm{t})\tau(s', s, \mathrm{t}) + \delta(s, \mathrm{t})\left(1 - \sum_{s' \in \mathcal{S} \setminus \{s\}} \tau(s, s', \mathrm{t})\right)\end{aligned}$$

Cancelling the term $\delta(s, \mathrm{t})$ on both sides gives

$$\sum_{s' \in \mathcal{S} \setminus \{s\}} (\delta(s', \mathrm{t})\tau(s', s, \mathrm{t}) - \delta(s, \mathrm{t})\tau(s, s', \mathrm{t})) = 0.$$

Obviously, $\delta(s, \mathrm{t})\tau(s, s, \mathrm{t}) - \delta(s, \mathrm{t})\tau(s, s, \mathrm{t}) = 0$, and so we can add this term to the sum and obtain for each $s \in \mathcal{S}$ what is called the *balance equation*:

$$\sum_{s' \in \mathcal{S}} (\delta(s', \mathrm{t})\tau(s', s, \mathrm{t}) - \delta(s, \mathrm{t})\tau(s, s', \mathrm{t})) = 0. \tag{3.11}$$

3.4 Reversible chains

The balance equation is certainly satisfied if all the individual terms vanish. Since δ is strictly positive, this can only happen when $\tau(s', s, \mathrm{t}) \neq 0$ whenever $\tau(s, s', \mathrm{t}) \neq 0$, and this in turn requires μ to be a symmetric relation, i.e.

$$\forall_{s' \in \mathcal{S}} \forall_{s' \in \mathcal{S}} [(s, s') \in \mu \Rightarrow (s', s) \in \mu]. \tag{3.12}$$

Definition 3.3 A chain that satisfies the *detailed balance conditions*:

$$\forall_{s \in \mathcal{S}} \forall_{s' \in \mathcal{S}} [\delta(s, \mathrm{t})\tau(s, s', \mathrm{t}) = \delta(s', \mathrm{t})\tau(s', s, \mathrm{t})] \tag{3.13}$$

is called *reversible*.

■

The detailed balance conditions imply that all the individual terms in the balance equations are zero. In that case

$$\forall_{(s, s') \in \mu} \left[\frac{\delta(s', \mathrm{t})}{\delta(s, \mathrm{t})} = \frac{\alpha(\varepsilon(s), \varepsilon(s'), \mathrm{t})}{\alpha(\varepsilon(s'), \varepsilon(s), \mathrm{t})} \frac{\beta(s, s')}{\beta(s', s)}\right]. \tag{3.14}$$

This means that the equilibrium density of a reversible chain would be independent from the move set and the selection function if β is such that

$$\forall_{(s, s') \in \mu} [\beta(s, s') = \beta(s', s)], \tag{3.15}$$

and the requirements (3.2), (3.3), and (3.4) are not violated.

If there is a strictly positive function $\omega : \mathcal{S} \to \mathbb{R}_+$ for a given reflexive chain such that

$$\forall_{s \in \mathcal{S}} \forall_{s' \in \mathcal{S}} [\omega(s)\tau(s, s', t) = \omega(s')\tau(s', s, t)]$$

then summing over all $s \in \mathcal{S}$ shows that

$$\forall_{s \in \mathcal{S}} [\omega(s) = \sum_{s' \in \mathcal{S}} \omega(s')\tau(s', s, t)],$$

which shows that ω has to be a multiple of δ, and that the chain has to be reversible.

Theorem 3.6 A reflexive chain is reversible fif there exists a strictly positive function $\omega : \mathcal{S} \to \mathbb{R}_+$ such that

$$\forall_{s \in \mathcal{S}} \forall_{s' \in \mathcal{S}} [\omega(s)\tau(s, s', t) = \omega(s')\tau(s', s, t)].$$

That function is a multiple of the equilibrium density δ of the chain.

■

The reason for choosing the word reversible for these chains is that the statistics of such a chain in equilibrium are invariant under a sequence reversal about an arbitrary step. For assume that the detailed balance conditions are satisfied by the chain under consideration, then

$$\prod_{i=1}^{n-1} \delta(s_{q(i)}, t)\tau(s_{q(i)}, s_{q(i+1)}, t) = \prod_{i=1}^{n-1} \delta(s_{q(i+1)}, t)\tau(s_{q(i+1)}, s_{q(i)}, t).$$

If we divide both sides by $\prod_{i=2}^{n-1} \delta(s_{q(i)}, t)$, we obtain on the left hand side

$$\delta(s_{q(1)}, t) \prod_{i=1}^{n-1} \tau(s_{q(i)}, s_{q(i+1)}, t),$$

which is the probability that at steps $N, N+1, N+2, \ldots, N+n-1$ the current state is respectively $s_{q(1)}, s_{q(2)}, \ldots, s_{q(n)}$ and the right hand side

$$\delta(s_{q(n)}, t) \prod_{i=1}^{n-1} \tau(s_{q(i+1)}, s_{q(i)}, t).$$

which is the probability that at steps $N', N'-1, N'-2, \ldots, N'-n+1$ the current state is respectively $s_{q(1)}, s_{q(2)}, \ldots, s_{q(n)}$.

The equality of these two probabilities for all families $\{s_{q(1)}, s_{q(2)}, \ldots, s_{q(n)}\}$ is the generally accepted definition of reversibility for stochastic processes. A straightforward specialization for $n = 2$ shows that any chain satisfying the general definition of reversability also has to satisfy the detailed balance conditions. Consequently, satisfying the detailed balance conditions is necessary and sufficient for a chain to be reversible in equilibrium in the above sense.

Clearly, whether a chain is reversible depends completely on the transition probabilities, and it must, therefore, be possible to express the conditions for reversibility in terms of transition probabilities only. These expressions can have a nice and compact form as is the case in the so-called *kolmogorov conditions*, that require for any subfamily $\{s_{q(1)}, s_{q(2)}, \ldots, s_{q(p)}\}$ of $\mathcal{S}$ with $q(1) = q(p)$.

$$\prod_{i=1}^{p} \tau(s_{q(i)}, s_{q(i+1)}, t) = \prod_{i=1}^{p} \tau(s_{q(i+1)}, s_{q(i)}, t).$$

We will prove that they are equivalent to the detailed balance conditions. But before passing on to that proof, note what is stated by these conditions: for any state the probability that the chain returns to that state via a given sequence of states is equal to the probability that it returns to that state via the same states in reversed sequence.

In order to prove the equivalence we rewrite the detailed balance equation in terms of matrices:

Theorem 3.7 A reflexive chain with $\mathbf{T}$ as its transition matrix and $\mathbf{d}$ as the left eigenvector of $\mathbf{T}$ for eigenvalue 1, is *reversible* fif

$$\mathbf{DT} = \mathbf{T}^{\mathsf{T}}\mathbf{D} \tag{3.16}$$

where $\mathbf{D}$ is the diagonal matrix with $d_{ii} = d_i$.

■

Theorem 3.8 A reflexive chain with transition matrix $\mathbf{T}$ is reversible fif for any subfamily of indices $Q = \{q(1), q(2), \ldots, q(p)\}$

$$t_{q(p), q(1)} \prod_{i=1}^{p-1} t_{q(i), q(i+1)} = t_{q(1), q(p)} \prod_{i=1}^{p-1} t_{q(i+1), q(i)}$$

■

Proof: If the chain is reversible then

$$\forall_{1 \leq i \leq s} \forall_{1 \leq j \leq s} [d_i t_{ij} = d_j t_{ji}],$$

and therefore, by multiplying all left hand side members with subsequent indices in Q and all right hand side members with subsequent indices in Q, (by subsequent we mean subsequent module p in the argument of q) we obtain

$$d_{q(p)} t_{q(p),q(1)} \prod_{i=1}^{p-1} d_{q(i)} t_{q(i),q(i+1)} = d_{q(1)} t_{q(1),q(p)} \prod_{i=1}^{p-1} d_{q(i+1)} t_{q(i+1),q(i)}.$$

$$t_{q(p),q(1)} \prod_{i=1}^{p} d_{q(i)} \prod_{i=1}^{p-1} t_{q(i),q(i+1)} = t_{q(1),q(p)} \prod_{i=1}^{p} d_{q(i)} \prod_{i=1}^{p-1} t_{q(i+1),q(i)}$$

$$t_{q(p),q(1)} \prod_{i=1}^{p-1} t_{q(i),q(i+1)} = t_{q(1),q(p)} \prod_{i=1}^{p-1} t_{q(i+1),q(i)}.$$

Conversely, if the kolmogorov conditions are satisfied, then the terms on both sides in

$$t_{q(p),q(1)} \sum_{q(2)=1}^{n} \sum_{q(3)=1}^{n} \cdots \sum_{q(p-1)=1}^{n} \prod_{i=1}^{p-1} t_{q(i),q(i+1)} =$$

$$= t_{q(1),q(p)} \sum_{q(2)=1}^{n} \sum_{q(3)=1}^{n} \cdots \sum_{q(p-1)=1}^{n} \prod_{i=1}^{p-1} t_{q(i+1),q(i)}$$

are one by one equal. This equality, however, can also be written as

$$t_{q(p),q(1)} \mathbf{U}_{q(1),q(p)} = t_{q(1),q(p)} \mathbf{U}_{q(p),q(1)}$$

with $\mathbf{U} = \mathbf{T}^p$. For $p \to \infty$ we obtain

$$t_{q(p),q(1)} \mathbf{T}^{\infty}_{q(1),q(p)} = t_{q(1),q(p)} \mathbf{T}^{\infty}_{q(p),q(1)}$$

which according to theorem 3.4 is equivalent to

$$t_{q(p),q(1)} d_{q(p)} = t_{q(1),q(p)} d_{q(1)}.$$

qed

We conclude this chapter with a surprising property of transition matrices of reversible chains.

Theorem 3.9 A reflexive chain with transition matrix **T** is reversible fif its transition matrix is the product of a diagonal matrix and a symmetric matrix.

■

Proof: Since $\mathbf{D}^\top = \mathbf{D}$, we have for a reversible chain that $\mathbf{DT} = \mathbf{T}^\top\mathbf{D} = \mathbf{T}^\top\mathbf{D}^\top = (\mathbf{DT})^\top$, or equivalently that $\mathbf{DT}$ is a symmetric matrix. Hence, $\mathbf{D}^{-1}(\mathbf{DT})$ is the product of a diagonal matrix and a symmetric matrix. Conversely, if $\mathbf{T} = \mathbf{BA}$ with $\mathbf{A}^\top = \mathbf{A}$ and $\mathbf{B}$ is a diagonal matrix, then $\mathbf{B}$ cannot have a zero on the diagonal and $\mathbf{A} = \mathbf{B}^{-1}\mathbf{T} = (\mathbf{B}^{-1}\mathbf{T})^\top = \mathbf{T}^\top\mathbf{B}^{-1}$. Without loss of generality it can be assumed that the sum of the reciprocals of the diagonal entries of $\mathbf{B}$ equals unity. From $\mathbf{B}^{-1}\mathbf{T}\mathbf{j} = \mathbf{T}^\top\mathbf{B}^{-1}\mathbf{j}$, it easily follows that $b_{ii} = d_i^{-1}$. So, $\mathbf{B}^{-1}\mathbf{T} = \mathbf{T}^\top\mathbf{B}^{-1}$ are the detailed balance equations.

qed

Corollary 3.1 The eigenvalues of the transition matrix of a reversible chain are all real.

■

Proof: If we denote by $\mathbf{D}^{\frac{1}{2}}$ the diagonal matrix with $\sqrt{d_i}$ on the diagonal, then

$$\mathbf{D}^{\frac{1}{2}}\mathbf{T}\mathbf{D}^{-\frac{1}{2}} = \mathbf{D}^{-\frac{1}{2}}(\mathbf{DT})\mathbf{D}^{-\frac{1}{2}}.$$

This means that **T** is similar to a symmetric matrix, and therefore has the same spectrum as this symmetric matrix. Since all eigenvalues of a symmetric matrix are real, all eigenvalues of **T** must be real.

qed

3.5 Discussion

The central result of this chapter, the chain limit theorem, is here obtained surprisingly by very elementary methods. Of course, the fact that $\mathcal{S}$ is finite allows us to avoid the complications of the proofs of Erdös Feller and Pollard [34] or Kolmogorov [86]. We also did not have to invoke the Perron-Frobenius theorem[41], because our transition matrices are very special nonnegative matrices. And the pay-off does not end here, as we shall see in later chapters.

The basic reversibility results are due to Kolmogorov[86],[85]. The observation that the transition matrix of a reversible chain has exclusively real eigenvalues, was not known to us from literature. Yet, once this fact was suggested by experiments the proof turned out to be quite simple.

4 CHAIN STATISTICS

Chains as introduced in the previous chapter are stochastic processes, because selection and acceptance functions assign probabilities to state pairs. Probability theory is therefore another basic discipline needed for deriving and controlling the properties of chains. In many of these manipulations a key role is played by the equilibrium density function. We want to derive properties of this function from the properties of selection and acceptance function, and we want to translate requirements on the equilibrium density function into constraints on the selection and acceptance function. These constraints do not determine the equilibrium density function completely. The score function also has a major influence, and is derived from the instance of the optimization problem at hand. In typical applications the number of states is too large to determine the equilibrium density function exactly. We therefore have to estimate the properties of this function by observing state and score frequencies. This falls in the domain of statistics.

The purpose of this chapter is to present notations and elementary facts from the two disciplines, probability theory and statistics. Most of it should be familiar to the reader, certainly the first three sections on density functions, expectations and sampling. Section 4.4 is a preparation for finding the most likely density function that satisfies given expectations. An unusual concept is perhaps the accessibility of a state space. The last section of this chapter is devoted to defining and discussing that concept. Although the state spaces considered in this book are finite and thus lead to discrete ensembles, this chapter deals predominantly with continuous distributions. The reason for this is that continuous distributions will be used extensively, and it is the more general case.

4.1 Density Functions

A *random variable* is a (real) number assigned to the outcome of an experiment. n random variables associated with an experiment can form a random vector variable $\mathbf{x}_n$. Since these random variables assume a value for every outcome, it is natural to consider events such as $\mathbf{x} \in D \subseteq \mathbb{R}^n$ and its probability $P(\mathbf{x} \in D \subseteq \mathbb{R}^n)$. (The *probability* $P(\mathbf{x} \in D \subseteq \mathbb{R}^n)$ where $D = \{u | c(u)\}$ is defined by a condition $c : \mathbb{R}^n \to \{0,1\}$, is loosely written as $P(c(\mathbf{x}))$.) For this purpose the (*n-variate*) *distribution function* associated with $\mathbf{x}$ is defined as the function F:

$$F(\mathbf{w}) = P(x_1 \leq w_1, \ldots, x_n \leq w_n) = P(\mathbf{x} \leq \mathbf{w}).$$

This function is non-decreasing in its arguments, right-continuous, and satisfies, $\lim_{w_i \to -\infty} F(\mathbf{w}) = 0$ and $\lim_{\mathbf{w} \to +\infty} F(\mathbf{w}) = 1$. The probability is strictly speaking a Lebesgue-Stieltjes integral usually written as

$$\int_D dF(\mathbf{x}).$$

Although we will use this notation from time to time, the subtleties behind it will be ignored.

In the case of discrete ensembles the distribution functions are sums of step-functions. In the continuous case there exists a non-negative Lebesgue-measurable function $f(\mathbf{x})$ such that

$$F(\mathbf{x}) = \int_{-\infty}^{x_n} \cdots \int_{-\infty}^{x_1} f(\mathbf{x}) dx_1 \ldots dx_n$$

wherever $\frac{\partial^n F(\mathbf{x})}{\partial x_1 \ldots \partial x_n}$ exists and $\frac{\partial^n F(\mathbf{x})}{\partial x_1 \ldots \partial x_n} = f(\mathbf{x})$ except possibly for a set with probability 0. $f(\mathbf{x})$ is called the *probability density function*. (More precisely, such a density function $f(\mathbf{x})$ exists fif $F(\mathbf{x})$ is absolutely continuous, and a random variable having such a distribution function is called an *absolutely continuous random variable*.) For discrete ensembles the total probability density is concentrated in a countable set of points.

The following terminology is used to describe densities and distributions of partitioned vectors. Consider the partitioned vector $\mathbf{x}^T = (\mathbf{y}^T, \mathbf{z}^T)$, where $\mathbf{y} \in \mathbb{R}^k$ and $\mathbf{z} \in \mathbb{R}^{n-k} (1 < k < n)$. Let $\mathbf{x}$ have density function $f(\mathbf{x})$ and distribution $F(\mathbf{x})$. $F(\mathbf{x}) = F(\mathbf{y}, \mathbf{z})$ may then be described as the *joint distribution function* of $\mathbf{y}$ and $\mathbf{z}$ and $f(\mathbf{y}, \mathbf{z})$ is called the *joint density function*. Further, the function

$$P(\mathbf{y} \leq \mathbf{v}) = F(v_1, \ldots, v_k, \infty, \ldots, \infty)$$

with vectors $\mathbf{v}$ from $\mathbb{R}^k$ as arguments, is called the *marginal distribution function* of $\mathbf{y}$. The *marginal density function* $g(\mathbf{y})$ of $\mathbf{y}$ is given by

$$g(\mathbf{y}) = \int_{\mathbb{R}^{n-k}} f(\mathbf{y}, \mathbf{z}) dz_1 \ldots dz_{n-k}.$$

In the previous chapter conditional probabilities were already mentioned. If $f(\mathbf{x})$ is continuous and the marginal density function $g(\mathbf{y})$ does not vanish it is safe to write for the conditional probability

$$P(\mathbf{z} \leq \mathbf{w} | \mathbf{v} < \mathbf{y} \leq \mathbf{v} + \mathbf{h}) = \frac{\int_{\mathbf{v}}^{\mathbf{v}+\mathbf{h}} \int_{-\infty}^{\mathbf{w}} f(\mathbf{y}, \mathbf{z}) dz_1 \ldots dz_{n-k} dy_1 \ldots dy_k}{\int_{\mathbf{v}}^{\mathbf{v}+\mathbf{h}} g(\mathbf{y}) dy_1 \ldots dy_k}.$$

Dividing numerator and denominator by $|\mathbf{h}|$ and letting $\mathbf{h} \to \mathbf{o}$ shows that the right hand side tends to

$$\frac{\int_{-\infty}^{\mathbf{w}} f(\mathbf{v}, \mathbf{z}) dz_1 \ldots dz_{n-k}}{g(\mathbf{v})}.$$

For fixed $\mathbf{v}$ this is a distribution function in $\mathbf{w}$ with as a density function

$$\frac{f(\mathbf{v}, \mathbf{w})}{g(\mathbf{v})}.$$

The *conditional density function* of $\mathbf{z}$ given the value of $\mathbf{y}$ is therefore defined as

$$c(\mathbf{z} \,|\, \mathbf{y}) = \frac{f(\mathbf{y}, \mathbf{z})}{g(\mathbf{y})}.$$

When the conditional density function $c(\mathbf{z}|\mathbf{y} = \mathbf{v})$ is the same for all values of $\mathbf{v}$, then we say that $\mathbf{y}$ and $\mathbf{z}$ are *statistically independent* of each other. In such situations, $c(\mathbf{z}|\mathbf{y} = \mathbf{v}) = h(\mathbf{z})$, the density function of $\mathbf{z}$. Hence, $\mathbf{y}$ and $\mathbf{z}$ are statistically independent fif $f(\mathbf{x}) = g(\mathbf{y})h(\mathbf{z})$ for all $\mathbf{x}$.

In the introduction of this chapter we mentioned that we want to draw conclusions from observing chains. On a smaller scale this means that we are interested in what we can say about the occurrence of a certain event if we have observed another event. For purposes like that another quantity, called *information*, is introduced. It measures the probability that event B occurred when event A has been observed, relative to the probability that B occurred regardless of any observation:

$$I(A, B) = \ln\left(\frac{P(B \,|\, A)}{P(B)}\right).$$

With the identity $P(A, B) = P(B \,|\, A)P(A)$ when $P(A) \neq 0$ we see that

$$I(A, B) = \ln\left(\frac{P(B \,|\, A)}{P(B)}\right) = \ln\left(\frac{P(A, B)}{P(A)P(B)}\right) = \ln\left(\frac{P(A \,|\, B)}{P(A)}\right).$$

So, interchanging A and B does not affect the value of I. It is therefore called *mutual information.* The definition for the continuous case is in terms of densities (and can be obtained from the above definition as a limit case if the respective limits exist)

$$i(y, z) = \ln\left(\frac{c(z \mid y)}{g(z)}\right) = \ln\left(\frac{f(y, z)}{h(y)g(z)}\right).$$

We conclude this section with the changing of variables. Let $\mathbf{x}$ be a random variable with density function $f(\mathbf{x})$. Suppose that $\mathbf{y}_n$ is a random variable resulting from a transformation $\mathbf{u}$ from the random variable $\mathbf{x}$ with density function $f(\mathbf{x})$, to y: $\mathbf{y} = \mathbf{u}(\mathbf{x})$. ($\mathbf{u}$ has to satisfy certain mild conditions to make $\mathbf{y}$ also a random variable.) There is, of course, a strong relation between the density function of $\mathbf{x}$ and the density function $f'(\mathbf{y})$ of $\mathbf{y}$, and sometimes $f'(\mathbf{y})$ can be derived from $F(\mathbf{x})$. With any open region S where each component function of the vector function $\mathbf{u}$ has a unique inverse and where it possesses continuous first derivatives such that the Jacobian $J = \det(\left[\frac{\partial x_i}{\partial y_j}\right]) \neq 0$, there corresponds an image S′ in which at any point $\mathbf{y}$ we have

$$f'(\mathbf{y}) = f(\mathbf{u}^{-1}(\mathbf{y})) \cdot |J|$$

and

$$\int_S f(\mathbf{x})dx_1 \ldots dx_n = \int_{S'} f(\mathbf{u}^{-1}(\mathbf{y})) \cdot |J| dy_1 \ldots dy_n.$$

4.2 Expected values

If $\mathbf{x}$ is a random vector variable with density function $f(\mathbf{x})$ then the *expectation* of a scalar valued function $\alpha(\mathbf{x})$ is defined as

$$\langle \alpha(\mathbf{x}) \rangle = \int_{\mathbb{R}^n} \alpha(\mathbf{x}) dF(\mathbf{x}).$$

More generally, the expectation of a matrix valued (or vector valued) function of $\mathbf{x}$, $\mathbf{A}(\mathbf{x})$ in which each entry $\alpha_{ij}(\mathbf{x})$ is a function of $\mathbf{x}$, is defined to be the matrix $\langle \mathbf{A}(\mathbf{x}) \rangle$ in which each entry is $\langle \alpha_{ij}(\mathbf{x}) \rangle$.

Assuming that all necessary integrals converge, we have the following linearity property

$$\langle a\alpha(\mathbf{x}) + b\beta(\mathbf{x}) \rangle = a\langle \alpha(\mathbf{x}) \rangle + b\langle \beta(\mathbf{x}) \rangle$$

If $\mathbf{y}$ and $\mathbf{z}$ are independent and $\alpha(\mathbf{y})$ is a function of $\mathbf{y}$ only and $\beta(\mathbf{z})$ is a function of $\mathbf{z}$ only, then

$$\langle \alpha(\mathbf{y})\beta(\mathbf{z}) \rangle = \langle \alpha(\mathbf{y}) \rangle \langle \beta(\mathbf{z}) \rangle$$

The vector $\langle \mathbf{x} \rangle = \mu$ is called the *mean* or the *mean vector* of $\mathbf{x}$. Thus,

$$\mu_i = \int_{\mathbb{R}^n} x_i \, dF(\mathbf{x}) = \int_{-\infty}^{+\infty} x_i dF_i(x_i)$$

where $F_i(x_i)$ is the marginal distribution of x_i. Of course, from the linearity property for expectations it follows that the mean vector has the following linearity property

$$\langle \mathbf{A}_{m,n} \mathbf{x} + \mathbf{b}_m \rangle = \mathbf{A}_{m,n} \mu + \mathbf{b}_m$$

when $\mathbf{A}$ and $\mathbf{b}$ do not depend on $\mathbf{x}$.

The *covariance* between two vectors, $\mathbf{x}_p$ and $\mathbf{y}_q$, defined as the matrix

$$\mathrm{Cov}(\mathbf{x}, \mathbf{y}) = \langle (\mathbf{x} - \mu)(\mathbf{y} - \nu)^T \rangle$$

where $\mu = \langle \mathbf{x} \rangle$, $\nu = \langle \mathbf{y} \rangle$. This matrix has row order p and column order q. The matrix

$$\mathrm{Cov}(\mathbf{x}, \mathbf{x}) = \langle (\mathbf{x} - \mu)(\mathbf{x} - \mu)^T \rangle =: \mathbf{S}$$

is called the *covariance matrix* of $\mathbf{x}$. The covariance of a random scalar variable x is a non-negative scalar $\langle (x - \mu)^2 \rangle$ called *variance*. The variance is denoted by σ^2, and its positive square root σ is called the *standard deviation*. Notice the following simple properties of covariances.

$$\mathbf{S} = \langle \mathbf{x}\mathbf{x}^T \rangle - \mu\mu^T$$

$$\mathrm{Cov}(\mathbf{A}\mathbf{x} + \mathbf{b}, \mathbf{C}\mathbf{x} + \mathbf{d}) = \mathbf{A}\,\mathrm{Cov}(\mathbf{x}, \mathbf{x})\mathbf{C}^T$$

$$\mathrm{Cov}(\mathbf{x}, \mathbf{y}) = \mathrm{Cov}(\mathbf{y}, \mathbf{x})^T$$

$$\mathrm{Cov}(\mathbf{x} + \mathbf{y}, \mathbf{x} + \mathbf{y}) = \mathrm{Cov}(\mathbf{x}, \mathbf{x}) + \mathrm{Cov}(\mathbf{x}, \mathbf{y}) + \mathrm{Cov}(\mathbf{y}, \mathbf{x}) + \mathrm{Cov}(\mathbf{y}, \mathbf{y})$$

the latter if $\mathbf{x}$ and $\mathbf{y}$ have an equal number of components. If $\mathbf{x}$ and $\mathbf{y}$ are independent then Cov($\mathbf{x}$,$\mathbf{y}$)=0. The converse, however, is not always true. For all constant vectors $\mathbf{a}$

$$\mathrm{Cov}(\mathbf{a}^T\mathbf{x}, \mathbf{a}^T\mathbf{x}) = \mathbf{a}^T \mathrm{Cov}(\mathbf{x}, \mathbf{x})\mathbf{a} = \mathbf{a}^T\mathbf{S}\mathbf{a} \tag{4.1}$$

Since the left-hand side of (4.1) is the variance of the scalar $\mathbf{a}^T\mathbf{x}$, which is always non-negative, it follows that $\mathbf{S}$ is always positive semidefinite.

Mutual information, defined in the section 4.1 is a random variable, and it has of course an expected value just like any other random variable. This expected value is sometimes called the *trans-information* of two ensembles. It turns out to be the difference of two other interesting expectations:

$$\langle i(\mathbf{y},\mathbf{z})\rangle = \int_{\mathbb{R}^n} f(\mathbf{y},\mathbf{z}) \ln\left(\frac{c(\mathbf{z}\mid\mathbf{y})}{h(\mathbf{z})}\right) dy_1 \ldots dy_k dz_1 \ldots dz_{n-k} =$$

$$= \int_{\mathbb{R}^n} f(\mathbf{y},\mathbf{z}) \ln\left(c(\mathbf{z}\mid\mathbf{y})\right) dy_1 \ldots dy_k\, dz_1 \ldots dz_{n-k} - \int_{\mathbb{R}^{n-k}} h(\mathbf{z}) \ln\left(h(\mathbf{z})\right) dz_1 \ldots dz_{n-k}.$$

(The function names c,f,g,and h correspond with those in section 4.1.) Such expectations are called entropies, and play a significant role in modern science. For a random vector variable $\mathbf{x}$ with density function $f(\mathbf{x})$ the *entropy* $\mathcal{H}$ of $\mathbf{x}$ is defined as

$$\mathcal{H}(\mathbf{x}) = -\langle \ln(f(\mathbf{x}))\rangle$$

If $\mathbf{x}$ is partitioned in two vectors $\mathbf{x}^T = (\mathbf{y}^T, \mathbf{z}^T)$ with joint density $f(\mathbf{y},\mathbf{z})$ then the *joint entropy* is defined as the entropy of the joint density function: $\mathcal{H}(\mathbf{y},\mathbf{z}) = \mathcal{H}(\mathbf{x})$. The *conditional entropy* is the entropy of the conditional density function:

$$\mathcal{H}(\mathbf{z}|\mathbf{y}) = \langle -\ln(c(\mathbf{z}|\mathbf{y}))\rangle = \langle -\ln\left(\frac{f(\mathbf{y},\mathbf{z})}{g(\mathbf{y})}\right)\rangle = \mathcal{H}(\mathbf{y},\mathbf{z}) - \mathcal{H}(\mathbf{y})$$

Using these definitions we can rewrite the trans-information as

$$\langle i(\mathbf{y},\mathbf{z})\rangle = \mathcal{H}(\mathbf{z}) - \mathcal{H}(\mathbf{z}\mid\mathbf{y}) = \mathcal{H}(\mathbf{y}) - \mathcal{H}(\mathbf{y}\mid\mathbf{z}) = \mathcal{H}(\mathbf{y}) + \mathcal{H}(\mathbf{z}) - \mathcal{H}(\mathbf{y},\mathbf{z})$$

For discrete ensembles with a finite number N of events it is quite easy to find bounds for the entropy:

$$0 \le \mathcal{H}(\mathbf{x}) \le \ln(|N|).$$

The lower bound is achieved when only one event is possible, and the upper bound when all events are equally probable. With some more effort one can show that the lower bound is also valid in the continuous case. However, no upper bound exists in general. The conditional entropy is bounded by $0 \le \mathcal{H}(\mathbf{z}|\mathbf{y}) \le \mathcal{H}(\mathbf{z})$ with the last sign being equal fif $\mathbf{y}$ and $\mathbf{z}$ are statistically independent. It follows that $\mathcal{H}(\mathbf{y},\mathbf{z}) \ge \mathcal{H}(\mathbf{y})$. The joint entropy of a vector of variables can be written as a series expansion of the conditional entropies of the components of the vector:

$$\mathcal{H}(\mathbf{x}) = \mathcal{H}(x_0) + \sum_{i=1}^{n} \mathcal{H}(x_i|x_{i-1} \ldots x_0). \tag{4.2}$$

4.3 Sampling

Often, when the density function of a random process is not known, the process is being observed to obtain some knowledge about the probability law governing the process. We will assume that a number of independent samples of the random process can be taken. The frequency data can be used as an estimate for the density function. Also estimates for the expectation of a function of the random variable can be calculated from this data. Often, such an *estimator* can be used in combination with prior information to determine the density function exactly.

Suppose $\mathbf{X}_{k,n}$ is a matrix whose columns are n independent samples from a k-variate distribution with mean μ and covariance matrix $\mathbf{S}$. The *average* of this sample is denoted by $\bar{\mathbf{x}}$, and defined as

$$\bar{\mathbf{x}} = \frac{1}{n}\mathbf{X}\mathbf{j} \tag{4.3}$$

A useful and easily verified property of this average is that for any vector $\mathbf{c} \in \mathbb{R}^n$

$$(\mathbf{X} - \mathbf{c}\mathbf{j}^T)(\mathbf{X} - \mathbf{c}\mathbf{j}^T)^T = (\mathbf{X} - \bar{\mathbf{x}}\mathbf{j}^T)(\mathbf{X} - \bar{\mathbf{x}}\mathbf{j}^T)^T + (\bar{\mathbf{x}} - \mathbf{c})(\bar{\mathbf{x}} - \mathbf{c})^T \tag{4.4}$$

$\bar{\mathbf{x}}$ is a random vector variable, of course, and we are often interested in its mean and covariance matrix. Let us first take the mean value of both sides of its definition:

$$\langle\bar{\mathbf{x}}\rangle = \langle\frac{1}{n}\mathbf{X}\mathbf{j}\rangle = \frac{1}{n}\sum_{i=1}^{n}\langle\mathbf{x}_{.,i}\rangle = \frac{1}{n}\sum_{i=1}^{n}\mu = \mu$$

$\bar{\mathbf{x}}$ is therefore said to be an *unbiased estimator* for μ.

Now consider the covariance matrix of $\bar{\mathbf{x}}$:

$$\mathrm{Cov}(\bar{\mathbf{x}},\bar{\mathbf{x}}) = \frac{1}{n^2}\mathrm{Cov}(\mathbf{X}\mathbf{j},\mathbf{X}\mathbf{j}) = \frac{1}{n^2}\sum_{i=1}^{n}\sum_{j=1}^{n}\mathrm{Cov}(\mathbf{x}_{.,i},\mathbf{x}_{.,j}) =$$

$$= \frac{1}{n^2}\sum_{i=1}^{n}\mathrm{Cov}(\mathbf{x}_{.,i},\mathbf{x}_{.,i}) = \frac{1}{n}\mathbf{S}$$

because $\mathrm{Cov}(\mathbf{x}_{.,i},\mathbf{x}_{.,j}) = 0$ whenever $i \neq j$. (Remember that the samples are independent.) That $\mathrm{Cov}(\bar{\mathbf{x}},\bar{\mathbf{x}})$ decreases with increasing n means that the difference $\bar{\mathbf{x}} - \mu$ decreases. $\bar{\mathbf{x}}$ becomes a more accurate estimator as the number of samples increases:

$$\lim_{n\to\infty}\bar{\mathbf{x}} = \mu$$

Another important object in sampling is the random matrix variable

$$\underline{\mathbf{S}} = \frac{1}{n-1}\mathbf{X}(\mathbf{I} - \frac{1}{n}\mathbf{J})\mathbf{X}^T \tag{4.5}$$

called the *sampling covariance matrix.*

Note that

$$(\mathbf{I} - \frac{1}{n}\mathbf{J})^T = \mathbf{I} - \frac{1}{n}\mathbf{J} = (\mathbf{I} - \frac{1}{n}\mathbf{J})^2,$$

and that therefore

$$\underline{\mathbf{S}} = \frac{1}{n-1}\mathbf{X}(\mathbf{I} - \frac{1}{n}\mathbf{J})\left(\mathbf{X}(\mathbf{I} - \frac{1}{n}\mathbf{J})\right)^T$$

which shows that $\underline{\mathbf{S}}$ is positive semidefinite. Also for this random variable we are interested in the mean value. (4.4) proves its usefulness in establishing the derivation. In

$$(n-1)\underline{\mathbf{S}} = \mathbf{X}(\mathbf{I} - \frac{1}{n}\mathbf{J})\mathbf{X}^T = \mathbf{X}\mathbf{X}^T - \frac{1}{n}(\mathbf{X}\mathbf{j})(\mathbf{X}\mathbf{j})^T = \mathbf{X}\mathbf{X}^T - n\,\bar{\mathbf{x}}\bar{\mathbf{x}}^T =$$
$$= (\mathbf{X} - \bar{\mathbf{x}}\mathbf{j}^T)(\mathbf{X} - \bar{\mathbf{x}}\mathbf{j}^T)^T = (\mathbf{X} - \boldsymbol{\mu}\mathbf{j}^T)(\mathbf{X} - \boldsymbol{\mu}\mathbf{j}^T)^T - n\,(\bar{\mathbf{x}} - \boldsymbol{\mu})(\bar{\mathbf{x}} - \boldsymbol{\mu})^T$$

is (4.4) used twice. To handle the expectation of this matrix we use the identity $\mathbf{A}_{k,n}\mathbf{A}_{k,n}^T = \sum_{h=1}^n \mathbf{a}_{.,h}(\mathbf{a}_{.,h})^T$:

$$(n-1)\langle\underline{\mathbf{S}}\rangle =$$

$$\sum_{h=1}^{n}\langle(\mathbf{x}_{.,h} - \boldsymbol{\mu})(\mathbf{x}_{.,h} - \boldsymbol{\mu})^T\rangle - n\langle(\bar{\mathbf{x}} - \boldsymbol{\mu})(\bar{\mathbf{x}} - \boldsymbol{\mu})^T\rangle$$

$$= n\mathbf{S} - \frac{1}{n}\langle\sum_{i=1}^{n}(\mathbf{x}_{.,i} - \boldsymbol{\mu})\sum_{j=1}^{n}(\mathbf{x}_{.,j} - \boldsymbol{\mu})^T\rangle =$$

$$= n\mathbf{S} - \frac{1}{n}\sum_{i=1}^{n}\sum_{j=1}^{n}\langle(\mathbf{x}_{.,i} - \boldsymbol{\mu})(\mathbf{x}_{.,j} - \boldsymbol{\mu})^T\rangle$$

This result was obtained using the linearity property of expectations. We can recognize the definition of covariance, and due to the fact that the samples are independent we obtain

$$\begin{aligned}(n-1)\langle\underline{\mathbf{S}}\rangle &= n\mathbf{S} - \tfrac{1}{n}\textstyle\sum_{i=1}^n\sum_{j=1}^n \mathrm{Cov}(\mathbf{x}_{.,i}, \mathbf{x}_{.,j})\\ &= n\mathbf{S} - \tfrac{1}{n}\textstyle\sum_{i=1}^n \mathrm{Cov}(\mathbf{x}_{.,i}, \mathbf{x}_{.,i}) = n\mathbf{S} - \mathbf{S}\end{aligned}$$

So, $\langle \underline{\mathbf{S}} \rangle = \mathbf{S}$, and therefore $\underline{\mathbf{S}}$ is an unbiased estimator for $\mathbf{S}$, which was the reason for using $n-1$ as the divisor in (4.5).

For univariate distribution the notations are usually slightly different. In that case we have n independent samples $x_1, x_2, \ldots, x_n$ and write

$$\bar{x} = \frac{1}{n}\sum_{i=1}^{n} x_i \quad \text{and} \quad s^2 = \frac{1}{n-1}\sum_{i=1}^{n}(x_i - \bar{x})^2. \tag{4.6}$$

4.4 Maximum likelyhood densities

Suppose that the only information we have about a certain experiment producing a random variable x with s possible outcomes are some expectations $\langle \rho_i(x) \rangle = R_i \quad (i = 0, 1, 2, \ldots, k)$. When the experiment is performed N times and outcome x_i occurs n_i times, what is then the most likely probability law behind the process? This, of course, depends on how "most likely" is defined. We have chosen as the "most likely" density function the one that allows the most ways of satisfying the constraints that were given as the expectations $\langle \rho_i(x) \rangle = R_i \quad (i = 0, 1, 2, \ldots, k)$. This turns out to be the density function that, in addition to satisfying the constraints, maximizes the entropy.

The number of possible sequences yielding the relative frequencies $\frac{n_i}{N}$ for each possible outcome is

$$W = \frac{N!}{n_1! n_2! n_3! \ldots n_s!}$$

Instead of maximizing W we may maximize any monotonic increasing function of W. $\frac{\ln(W)}{N}$ is such a function. Using Stirling's approximation in the form $\ln(n!) \approx n\ln(n) - n + \frac{1}{2}\ln(2\pi n)$ and taking the limit for $N \to \infty$ leads to

$$\lim_{N\to\infty} \frac{\ln(W)}{N} =$$

$$= \lim_{N\to\infty} \frac{1}{N}\left(N\ln(N) - N + \tfrac{1}{2}\ln(2\pi N) - \sum_{i=1}^{k}(n_i\ln(n_i) - n_i + \tfrac{1}{2}\ln(2\pi n_i))\right) =$$

$$\lim_{N\to\infty} \sum_{i=1}^{k} -\frac{n_i}{N}\ln\left(\frac{n_i}{N}\right) = \mathcal{H}(x)$$

The problem of finding the most likely distribution becomes then maximizing $\mathcal{H}(x)$ subject to

$$\langle \rho_i(x) \rangle = R_i \quad i = 0, 1, 2, \ldots, k$$

This problem can be solved with the method of lagrange multipliers, a standard technique for constrained optimization problems.

4.5 Aggregate functions

The average score of the states produced by a chain is certainly an an interesting characteristic of a chain in equilibrium. A relatively low average indicates that it is mainly producing states that correspond with configurations of high quality. Of course, the average can never be lower than ε_0, the absolute lowest score in the state space and, in general, an unknown number. Therefore, indications on whether the average score is close to minimum have to be derived from other quantities. One such a quantity is the variance of the score. When the average score is close to ε_0 the variance cannot be large. High scores have to be sporadic, because many low scores have to compensate their contribution to the average. But in many other ways the variance will prove to be a valuable piece of information for analyzing and controlling the chains. Another interesting quantity would be a measure of size for the part of the state space the chain is most likely to be. It would also be a kind of uncertainty measure: is any state equally likely to be the current state or is the chain almost exclusively producing global minimum states.

For these quantities we propose expected values that can be observed from a number of identical chains, or, for ergodic chains, from a single chain in action. The first quantity, the *average score*, is

$$E(t) = \langle \varepsilon \rangle = \sum_{s \in \mathcal{S}} \delta(s,t)\,\varepsilon(s) = \sum_{\{\varepsilon\}} \varepsilon\,\hat{\delta}(\varepsilon,t).$$

For a chain in equilibrium this will be a constant, so that for the *score variance* we can write

$$\sigma^2(t) = \langle (\varepsilon - E(t))^2 \rangle = \left| \langle \varepsilon^2 \rangle - E^2(t) \right|.$$

From section 4.3 and 3.4 we know that we can estimate $E(t)$ by $\bar{\varepsilon}$ of the successive states, that is by summing the scores of a state sequence and dividing them by the number of states in the sequence. And also from these sections we know that an unbiased estimator for the score variance is given by

$$\frac{1}{n-1} \sum_{i=1}^{n} (\varepsilon_i - \bar{\varepsilon})^2.$$

Then the third quantity, the one that is to measure in how big a subspace most of the visited states are. We will call the quantity the *accessibility* of the state space for the chain in equilibrium. This accessibility has to have certain properties. First, it should only depend on the relative frequency of the states, and thus on the equilibrium density δ. The frequency of a state depends on its score, of course, but states with equal scores should have identical effects on the accessibility. We therefore want the accessibility to be a function with the state frequencies as arguments: $\mathsf{H_s} : [0,1]^{\mathsf{s}} \longrightarrow \mathbb{R}$. That function must be symmetrical in its arguments. It should also be continuous with respect to each of its arguments: small changes in the frequencies should cause small changes in the accessibility. When all states are equally probable the accessibilty should assume its maximum value:

$$\forall_{\delta}[\mathsf{H_s}(\delta) \leq \mathsf{H_s}(\frac{1}{\mathsf{s}}\mathsf{j_s})]$$

Further, the value of H should not depend on the description of the chain. Suppose the state space $\mathcal{S}$ is partitioned into p subspaces forming the partition $P = \{\mathcal{S}_1, \mathcal{S}_2, \ldots, \mathcal{S}_\mathsf{p}\}$. The markov chain is then equivalent to a composition of several chains, one indicating in which block the current state is, and p chains indicating which state in the selected block is the current state. The equilibrium densities of chains are respectively

$$\delta_P(\mathcal{S}_\mathsf{i},\mathsf{t}) = \sum_{s \in \mathcal{S}_\mathsf{i}} \delta(s,\mathsf{t}) \text{ and for } s \in \mathcal{S}_\mathsf{i} \quad \delta_{\mathcal{S}_\mathsf{i}}(s,\mathsf{t}) = \frac{\delta(s,\mathsf{t})}{\sum_{s\prime \in \mathcal{S}_\mathsf{i}} \delta(s\prime,\mathsf{t})}$$

The accessibility measure should not depend on the chosen view point, and therefore

$$\mathsf{H_s}(\delta) = \mathsf{H_p}(\delta_P) + \sum_{\mathsf{i}=1}^{\mathsf{p}} \delta_P(\mathcal{S}_\mathsf{i},\mathsf{t})\mathsf{H}_{|\mathcal{S}_\mathsf{i}|}(\delta_{\mathcal{S}_\mathsf{i}})$$

Finally, adding an impossible state to the state set should not change the accessibility function:

$$\mathsf{H_s}(\mathsf{d_1},\mathsf{d_2},\ldots,\mathsf{d_s}) = \mathsf{H_{s+1}}(\mathsf{d_1},\mathsf{d_2},\ldots,\mathsf{d_s},0)$$

Theorem 4.1 $H_s(\mathbf{d}_s) = -c\sum_{i=1}^{s} d_i \ln(d_i)$, where c is an arbitrary positive constant, are the only functions, $H_s : [0,1]^s \longrightarrow \mathbb{R}$, continuous and symmetric in the arguments, satisfying

$$\forall_{\mathbf{0}_s \leq \mathbf{d}_s \leq \mathbf{j}_s, \sum d_i = 1}[H_s(\mathbf{d}_s) \leq H_s(\frac{1}{s}\mathbf{j}_s)], \tag{4.7}$$

$$H(d_1, d_2, \ldots, d_s) = H(d_1, d_2, \ldots, d_s, 0), \tag{4.8}$$

$$\forall_{\{\mathbf{P}_{s,p} | \mathbf{P}\mathbf{j}_p = \mathbf{j}_s\}}[H_s(\mathbf{d}) = H_p(\tilde{\mathbf{d}}) + \sum_{i=1}^{p} \tilde{d}_i H_s(\mathbf{DPI}_{\cdot,i})] \tag{4.9}$$

where the entries of $\mathbf{P}$ are zero or one, $\mathbf{D}$ is the diagonal matrix with $\forall_{1 \leq i \leq s}[d_{ii} = d_i]$ and $\tilde{\mathbf{d}} = \mathbf{P}^T\mathbf{D}\mathbf{j}_s$.

■

Proof: First, the form of H for all states equally likely is established. Then, using that result, the theorem is proved for rational arguments. Finally, because of the continuity requirement, the result is extended to all real arguments.

Let m(s) be $H_s(\frac{1}{s}\mathbf{j}_s)$. By (4.7) and (4.8):

$$m(s) = H_s(\frac{1}{s}\mathbf{j}_s) = H_{s+1}(\frac{1}{s}\mathbf{j}_s, 0) \leq H_{s+1}(\frac{1}{s+1}\mathbf{j}_{s+1}) = m(s+1) \tag{4.10}$$

With p and q two positive integers, let $\mathcal{S}$ be a state space with q^p states, partitioned in q subspaces $\{\mathcal{S}_1, \mathcal{S}_2, \ldots, \mathcal{S}_q\}$, each having q^{p-1} equally likely states. Then by (4.9)

$$m(q^p) =$$

$$= H_{q^p}(\frac{1}{q^p}\mathbf{j}_{q^p}) = H_q(\frac{1}{q}\mathbf{j}_q) + \sum_{i=1}^{q} \frac{1}{q} H_{q^{p-1}}(\frac{1}{q^{p-1}}\mathbf{j}_{q^{p-1}}) = m(q) + m(q^{p-1}),$$

and therefore

$$m(q^p) = p\, m(q). \tag{4.11}$$

Now consider two arbitrary fixed positive integers v and w, and take for q an integer greater than 1 with $m(q) \geq 0$ (assuming $m \geq 0$). Let p be such that $q^p \leq v^w < q^{p+1}$. Then,

$$\frac{p}{w} \leq \frac{\ln(v)}{\ln(q)} < \frac{p+1}{w} \tag{4.12}$$

Because of (4.10): $m(q^p) \leq m(v^w) \leq m(q^{p+1})$, and thus

$$\frac{p}{w} \leq \frac{m(v)}{m(q)} \leq \frac{p+1}{w} \tag{4.13}$$

By (4.12) and (4.13),

$$\left|\frac{m(v)}{m(q)} - \frac{\ln(v)}{\ln(q)}\right| \leq \frac{1}{w} \tag{4.14}$$

Since this must be valid for arbitrary large w, m(s) must be equal to $c\ln(s)$ with $c > 0$ because m is nondecreasing. The result obtained in this first part therefore is

$$H_s(\frac{1}{s}j_s) = c\ln(s). \tag{4.15}$$

Now suppose all d_i are rational and $\sum d_i = 1$. Then there exist s positive integers $g_1, g_2, \ldots, g_s$ such that

$$d_i = \frac{g_i}{g} \quad \text{where } g = \sum_{i=1}^{s} g_i. \tag{4.16}$$

Let G be a state space with g equally likely states, and P be a partition $\{\mathcal{G}_1, \mathcal{G}_2, \ldots, \mathcal{G}_s\}$ where $\mathcal{G}_i$ has g_i equally likely states. Then by (4.9), (4.15) and (4.16),

$$\begin{aligned} H_s(\mathbf{d}) &= H_g(\frac{1}{g}j_g) - \sum_{i=1}^{s}\left(\frac{g_i}{g}H_{g_i}(\frac{1}{g_i}j_{g_i})\right) = c\ln(g) - c\sum_{i=1}^{s}\left(\frac{g_i}{g}\ln(g_i)\right) = \\ &= c\ln(g) - c\sum_{i=1}^{s}\frac{g_i}{g}\left(\ln(g) + \ln(\frac{g_i}{g})\right) = \\ &= -c\sum_{i=1}^{s}\frac{g_i}{g}\ln(\frac{g_i}{g}) = -c\sum_{i=1}^{s} d_i\ln(d_i) \end{aligned} \tag{4.17}$$

Since H is assumed to be continuous, the theorem is proved for all density vectors **d** when we define $x\ln(x)$ as 0 for $x = 0$.

qed

The only functions that qualify for an accessibility function are (positive) multiples of the entropy of the equilibrium density function δ. Therefore, the entropy $\mathcal{H}(\delta) = H_t$ will be called the *accessibility* of a chain in equilibrium and denoted by the symbol H. It can also be seen as the uncertainty about the state that the process is in when only t (and a complete description of the state space) is known. For a given chain it only depends on t.

Since the annealing chain is a Markov process the conditional density of the state depends only on the previous state. Therefore also the conditional entropy depends only on the previous state:

$$\mathcal{H}(s_n|s_{n-1}\ldots s_0) = \mathcal{H}(s_n|s_{n-1})$$

This can be expressed in terms of the transition probability:

$$\mathcal{H}(s_n|s_{n-1}) = \mathcal{H}(s'|s) = \langle -\ln(\tau(s, s', t)) \rangle$$

This can be regarded as the amount of uncertainty about a single move of the annealing chain in equilibrium and will be referred to as the *local accessibility* $h_t = \mathcal{H}(s'|s)$.

The three expected values for a chain in equilibrium, the score average, the score variance and the space accessibility, will be called its *aggregate functions*. They will be used to characterize the chain while active on a particular state space. In chapter 5 we will see that these functions can be very useful in adapting the successive chains to the properties of the instance for which that state space was constructed.

4.6 Discussion

Most of this chapter is contained in any introductory into probability theory and statistics. The definition of maximum likelyhood is after [73], and is sometimes called *Jaynes' principle*. In the unconstrained case it reduces to the principle of insufficient information of Laplace which assigns equal probabilities to all possibilities. It is however a disputed principle, in spite of its success in numerous applications. The method was already used in [48] where the general solution was also obtained with the lagrange multiplier technique. The uniqueness proof for the accessibility function is from [80].

5 ANNEALING CHAINS

It will be clear from the previous chapters that once we have built the state space, and decided which selection function β and which acceptance function α is to be used, the equilibrium density δ is determined. If we make sure that the chain is reversible and β is symmetrical in its arguments, then δ only depends on α. So, if we want to enforce some desirable properties on this density we have to do that by selecting a proper acceptance function, and, when reversibility and symmetry are not implied, a proper move set and selection function as well. The realization of what seem to be desirable properties of a density function should be carefully considered. Some important aspects, such as the computational consequences for the implementation of the chain, are not always apparent from the density function. Also, what seems to be an advantage may not be under our control, or may already be inherent to chains. For example, a small score variance once we have obtained a chain that produces low average score, would be advantageous, because that would mean that the chain is almost all the time in a low score state, and thus selecting almost exclusively high quality configurations. Minimizing the score variance for all chains is not desirable, as will become clear in the later chapters, and if we are able to obtain chains with very low mean score, the variance has to be low as well, because the scores have a lower bound. What we certainly do not want is that the density function exhibits a bias towards certain states, that is not based on the score assigned to that state. States with equal scores are equally desirable, and should therefore have the same probability. This property is related to the accessibility of the chain.

5.1 Towards low scores

To be able to control the average score properly by setting t we require the acceptance function to be continuous and monotonic in t, a reasonable requirement considering that t is supposed to be a "control parameter". Further, we want to control that average in such a way that for small t this average will be low, that is, very close to ε_0. Although we can use the freedom left in specifying the move set, the selection function and the acceptance function for that purpose, it seems to be advantageous to do this through the acceptance function alone if possible, and to avoid additional constraints on the more problem related parts of the model, to wit μ and β. This might make it easier to fit actual optimization problems in the annealing framework. The constraints already imposed on the move set and the selection function are:

$$\bigcup_{k=1}^{\infty} \mu^k = \mathcal{S} \times \mathcal{S} \tag{3.1}$$

$$\forall_{(s,\, s') \notin \mu} \left[\beta(s,\, s') = 0\right] \tag{3.2}$$

$$\forall_{(s,\, s') \in \mu} \left[\beta(s,\, s') \neq 0\right] \tag{3.3}$$

$$\forall_{s \in \mathcal{S}} \left[\sum_{s' \in s\mu} \beta(s,\, s') = 1 \right]. \tag{3.4}$$

To these we add the requirement that the move set has to be a symmetric relation:

$$\forall_{s \in \mathcal{S}} \forall_{s' \in \mathcal{S}} \left[(s,\, s') \in \mu \Rightarrow (s',\, s) \in \mu\right]. \tag{3.12}$$

The mutual reachability in some finite number of moves is already implied in (3.1). Symmetry of the move set requires a state to be reachable in one move from another state, whenever it is reachable from that state in one move. This is an extra constraint, but is in practice almost always naturally achieved.

Now we want to find out how by choosing the acceptance function α we can effect that

$$\forall_{s \in \mathcal{S}} \left[\varepsilon(s) \neq \varepsilon_0 \Rightarrow \lim_{t \downarrow 0} \delta(s, t) = 0\right]. \tag{5.1}$$

For this purpose we give the states of $\mathcal{S}$ indices from $\{1,\ldots,s\}$ in such a way that

$$i<j \Rightarrow \varepsilon(s_i) \leq \varepsilon(s_j).$$

Using this indexing we form the matrix $\mathbf{B}_{ss}$ by putting $b_{ij} = \beta(s_i, s_j)$. Further we consider *state sequences* $(s_{q(0)}, s_{q(1)}, \ldots, s_{q(p)})$ that satisfy $\forall_{1\leq i\leq p}[(s_{q(i-1)}, s_{q(i)}) \in \mu]$. We distinguish *simple state sequences* in which no state occurs more than once, *cyclic state sequences* in which the first state is identical to the last state, and *complex state sequences.* A bijective relation ω over the set of state sequences associates with each state sequence a *return sequence* defined differently for the three types of state sequences. The return sequence of a simple state sequence has the same states, but in reverse order: $\omega(s_{q(0)}, s_{q(1)}, \ldots, s_{q(p)}) = (s_{q(p)}, s_{q(p-1)}, \ldots, s_{q(0)})$. For cyclic state sequences, where $s_{q(0)} = s_{q(p)}$ the return sequence is the same state sequence $\omega(s_{q(0)}, s_{q(1)}, \ldots, s_{q(p)}) = (s_{q(0)}, s_{q(1)}, \ldots, s_{q(p)})$. To define the return sequence for a complex state sequence we decompose that sequence into three subsequences. First, a simple state sequence $(s_{q(0)}, s_{q(1)}, \ldots, s_{q(i)})$ where $q(i)$ is the first index that occurs more than once in the original state sequence. Second, the longest cyclic subsequence with $q(i) = q(k)$. Finally, the third sequence with whatever remains: $(s_{q(k)}, s_{q(k+1)}, \ldots, s_{q(p)})$. After noting that all sequences with only two states are either simple or cyclic state sequences, and that the third subsequence certainly has fewer states than the original state sequence, it is clear that we can define the return sequence for complex state sequences inductively: $\omega(s_{q(0)}, s_{q(1)}, \ldots, s_{q(p)})$ is the concatenation of the return sequences of the three parts in reverse order: $\omega(s_{q(k)}, \ldots, s_{q(p)}) \circ (s_{q(i)}, s_{q(i+1)}, \ldots, s_{q(k)}) \circ (s_{q(i)}, s_{q(i-1)}, \ldots, s_{q(0)})$. To each state sequence two numbers are assigned:

$$w_B(s_{q(0)}, s_{q(1)}, \ldots, s_{q(p)}) = \prod_{i=1}^{p} b_{q(i-1),q(i)}$$

and

$$\rho_B = \frac{w_B(s_{q(0)}, s_{q(1)}, \ldots, s_{q(p)})}{w_B(\omega(s_{q(0)}, s_{q(1)}, \ldots, s_{q(p)}))}.$$

Lemma 5.1 For every non-cyclic state sequence there is a simple sequence with the same first state, the same last state, and the same ρ_B.

■

Proof: We use the same decomposition as in the definition of the complex state state sequence to establish the lemma by induction: the lemma is supposed to be true for all state sequences with less than p states. This implies that for $(s_{q(k)}, s_{q(k+1)}, \ldots, s_{q(p)})$ there is a $(s_{q'(0)}, s_{q'(1)}, \ldots, s_{q'(r)})$ with $\rho_B(s_{q(k)}, s_{q(k+1)}, \ldots, s_{q(p)}) = \rho_B(s_{q'(0)}, s_{q'(1)}, \ldots, s_{q'(r)})$, $q'(0) = q(k) = q(i)$, and $q'(r) = q(p)$.

$$\rho_B(s_{q(0)}, s_{q(1)}, \ldots, s_{q(p)}) =$$

$$= \frac{w_B(s_{q(0)}, \ldots, s_{q(i)}) w_B(s_{q(i)}, \ldots, s_{q(k)}) w_B(s_{q(k)}, \ldots, s_{q(p)})}{w_B(s_{q(i)}, \ldots, s_{q(0)}) w_B(s_{q(i)}, \ldots, s_{q(k)}) w_B(\omega(s_{q(k)}, \ldots, s_{q(p)}))} =$$

$$= \frac{w_B(s_{q(0)}, s_{q(1)}, \ldots, s_{q(i)}) w_B(s_{q(i)}, s_{q'(1)}, \ldots, s_{q'(r)})}{w_B(s_{q(i)}, s_{q(i-1)}, \ldots, s_{q(0)}) w_B(s_{q'(r)}, s_{q'(r-1)}, \ldots, s_{q(i)})} =$$

$$= \frac{w_B(s_{q(0)}, s_{q(1)}, \ldots, s_{q(i)}, s_{q'(1)}, \ldots, s_{q'(r)})}{w_B(s_{q'(r)}, s_{q'(r-1)}, \ldots, s_{q(i)}, s_{q(i-1)}, \ldots, s_{q(0)})} =$$

$$= \rho_B(s_{q(0)}, s_{q(1)}, \ldots, s_{q(k)}, s_{q'(1)}, \ldots, s_{q'(r)})$$

qed

Corollary 5.1 The ρ_B cannot exceed a finite $\max\{\rho_B\}$ which is the maximum over the ρ_B's of all simple state sequences.

■

Proof: Simple state sequences can have only finitely many states, and so their ρ_B's are finite, and there is a maximum > 1 among them. According to lemma 5.1 the ρ_B of a complex state sequence is equal to the ρ_B of some simple state sequence, and therefore cannot be greater than that maximum. All cyclic sequences have ρ_B equal to 1, and do not exceed the maximum either.

qed

B is not the state transition matrix $\mathbf{T}_{ss}$ of an annealing chain. In the present notation that matrix can be written as

$$t_{ij} = \begin{cases} b_{ij}a'_{ij} & \text{if } i > j \\ b_{ij}a_{ij} & \text{if } i < j \\ 1 - \sum_{\substack{k=1 \\ i\neq k}}^{s} t_{ik} & \text{if } i = j \end{cases}$$

Let us assume $\mathbf{A}' = \mathbf{J}_{ss}$, because the discussion is not essentially different with other admissible choices. Besides, it is in accordance with algorithm introduced in chapter 1. The ρ_T for simple state sequences and matrix **T** then becomes

$$\rho_T(s_{q(0)}, s_{q(1)}, \ldots, s_{q(p)}) = \frac{w_T(s_{q(0)}, s_{q(1)}, \ldots, s_{q(p)})}{w_T(\omega((s_{q(0)}, s_{q(1)}, \ldots, s_{q(p)}))} =$$

$$= \frac{w_B(s_{q(0)}, s_{q(1)}, \ldots, s_{q(p)})}{w_B(\omega(s_{q(0)}, s_{q(1)}, \ldots, s_{q(p)}))} \frac{\prod_{q(i-1)\geq q(i)} a_{q(i-1),q(i)}}{\prod_{q(i-1)\leq q(i)} a_{q(i),q(i-1)}} =$$

$$= \rho_B \frac{\prod_{q(i-1)\geq q(i)} a_{q(i-1),q(i)}}{\prod_{q(i-1)\leq q(i)} a_{q(i),q(i-1)}},$$

a rather unmanageable expression. But this changes when $\forall_{1\leq i\leq s}\forall_{1\leq j\leq s}\forall_{1\leq k\leq s}[a_{ij}a_{jk} = a_{ik}]$. This implies that $a_{ii} = 1$ and $a_{ij} = \frac{1}{a_{ji}}$. The expression is then valid for all state sequences, not only for simple state sequences, and becomes

$$\rho_T(s_{q(0)}, s_{q(1)}, \ldots, s_{q(p)}) =$$

$$= \rho_B(s_{q(0)}, s_{q(1)}, \ldots, s_{q(p)}) \prod_{i=1}^{p} a_{q(i),q(i-1)} =$$

$$= \rho_B(s_{q(0)}, s_{q(1)}, \ldots, s_{q(p)}) a_{q(0),q(p)}.$$

Consequently $\rho_T(s_{q(0)}, s_{q(1)}, \ldots, s_{q(p)}) \leq a_{q(0),q(p)} \max\{\rho_B\}$

Lemma 5.2 If $\mathbf{d}_s$ is the unique left eigenvector of **T** associated with the eigenvalue 1, then

$$\forall_{1\leq i<j\leq s}[d_j \leq a_{ij} \max\{\rho_B\}]$$

■

Proof: With $\mathbf{P} = \mathbf{T}^k$ the entry p_{ij} can be obtained by summing the w_T of all state sequences with k states from s_i to s_j, while the entry p_{ji} is equal to the sum of the w_T of all corresponding return sequences. Using the inequality $\rho_T(s_{q(0)}, s_{q(1)}, \ldots, s_{q(p)}) \leq a_{q(0),q(p)} \max\{\rho_B\}$ yields

$$p_{ij} \leq \frac{p_{ij}}{p_{ji}} \leq a_{ij} \max\{\rho_B\}.$$

Since $d_j = \lim_{k \to \infty} p_{ij}$, the lemma follows.

qed

This gives us a sufficient condition for achieving the goal formulated at the beginning of this section through the choice of the acceptance function.

Theorem 5.1 A reflexive chain with a symmetric move set has

$$\forall_{s \in \mathcal{S}}[\varepsilon(s) \neq \varepsilon_0 \Rightarrow \lim_{t \downarrow 0} \delta(s,t) = 0]$$

if it has an acceptance function α satisfying

$$\varepsilon \geq \varepsilon' \Rightarrow \alpha(\varepsilon, \varepsilon', t) = 1 \tag{5.2}$$

$$\varepsilon > \varepsilon' > \varepsilon'' \Rightarrow \alpha(\varepsilon, \varepsilon', t)\alpha(\varepsilon', \varepsilon'', t) = \alpha(\varepsilon, \varepsilon'', t) \tag{5.3}$$

$$\varepsilon \leq \varepsilon' \Rightarrow \lim_{t \downarrow 0} \alpha(\varepsilon, \varepsilon', t) = 0. \tag{5.4}$$

■

Note that no special constraints are imposed on β or μ, except for the symmetry of μ. This suggests that the state space can be freely constructed and that the desired properties of the successive chains can be achieved by choosing the right acceptance function. That there is more to it, will become clear in later chapters, but that can also be said of the more restrictive and yet more frequently used combinations of selection and acceptance.

It seems reasonable that smaller score increases are accepted with higher probability than larger ones, and that this probability varies smoothly with the score difference. This is elegantly achieved when $\alpha(\varepsilon, \varepsilon', t)$ is differentiable with respect to ε' and that the corresponding derivative is positive. Much stronger would be the requirement when, in addition, the acceptance may only depend on the score difference:

$$\alpha(\varepsilon, \varepsilon + \Delta\varepsilon, t) = \alpha(\varepsilon', \varepsilon' + \Delta\varepsilon, t) = f(\Delta\varepsilon, t) \tag{5.5}$$

With (5.3) this leads to

$$f(\Delta\varepsilon, t) = \frac{\alpha(\varepsilon_0, \varepsilon + \Delta\varepsilon, t)}{\alpha(\varepsilon_0, \varepsilon, t)}$$

Using the differentiability with respect to the second argument we obtain

$$\frac{\partial}{\partial\varepsilon}\alpha(\varepsilon_0, \varepsilon, t) = \lim_{\Delta\varepsilon \downarrow 0} \frac{\alpha(\varepsilon_0, \varepsilon + \Delta\varepsilon, t) - \alpha(\varepsilon_0, \varepsilon, t)}{\Delta\varepsilon} =$$

$$= \alpha(\varepsilon_0, \varepsilon, t) \lim_{\Delta\varepsilon \downarrow 0} \frac{f(\Delta\varepsilon, t) - f(0, t)}{\Delta\varepsilon} = \alpha(\varepsilon_0, \varepsilon, t)c(t).$$

This simple differential equation has as solution $\alpha(\varepsilon_0, \varepsilon, t) = k \exp(\varepsilon\, c(t))$.

Theorem 5.2 The only acceptance functions that are differentiable in the second argument, whose values depend on the third argument and the difference of the first two arguments, and that satisfy the conditions of theorem 5.1 are

$$\alpha(\varepsilon, \varepsilon', t) = \min\left\{1\,,\, \exp\left((\varepsilon' - \varepsilon)c(t)\right)\right\}.$$

where $c(t)$ has to be a negative, continuous and monotonic function with $\lim_{t \downarrow 0} c(t) = -\infty$.

■

Thus, by accepting strong constraints such as (5.3) and (5.5) the acceptance function is quite restricted in form. It should be noted however that neither constraint is necessary for achieving

$$\forall_{s \in S}\, [\varepsilon(s) \neq \varepsilon_0 \Rightarrow \lim_{t \downarrow 0} \delta(s, t) = 0].$$

If $\min\{1\,,\, \frac{\varepsilon}{\varepsilon'}\, t^{\varepsilon - \varepsilon'}\}$ were chosen, that goal would have been realized without satisfying (5.5). Also $\left(1 + \exp\left(\frac{\varepsilon - \varepsilon'}{t}\right)\right)^{-1}$ is adequate, but does not satisfy (5.3).

5.2 Maximal accessibility

The accessibility

$$H_s = \sum_{s \in \mathcal{S}} \delta(s, t) \ln(\delta(s, t))$$

depends on the values assigned to the s states by δ. The question we want to answer in this section is what values should δ assign to the states in order to make the accessibility as large as possible. Of course, these values have to satisfy the constraints

$$\forall_{s \in \mathcal{S}} [0 < \delta(s, t) < 1] \quad \text{and} \quad \sum_{s \in \mathcal{S}} \delta(s, t) = 1.$$

Maximizing the accessibility under these constraints alone will yield a familiar result. For in the previous chapter we saw that H was selected such that it attained the maximum value when all states were equally probable:

$$\forall_{s \in \mathcal{S}} [\, \delta(s, t) = \frac{1}{s} \,].$$

With such a δ the chain would produce random states as drawn from a uniform distribution. This clearly is not what we want to happen all of the time. Towards the end of the process we want the algorithm to produce the desirable states in overwhelming majority. And since the desirable states are characterized by a low score, an indication that a chain is producing predominantly desirable states, is that the average score over a fairly large number of steps is relatively low. So we would very much like to control that statistic. Let us therefore try to find what values δ should assign to the states to make the accessibility as large as possible for a given value E for the mean score. In the problem formulation this amounts to adding the constraint

$$\sum_{s \in \mathcal{S}} \varepsilon(s)\delta(s, t) = E.$$

Of course, setting E at a value higher than the highest score in the space, or lower than the lowest score in the space, is not realizable.

For convenience we give each state in S a unique index and introduce two vectors: **e** with $e_i = \varepsilon(s_i)$ and **d** with $d_i = \delta(s_i, t)$. Further, let $\ln(\mathbf{d})$ be the vector whose components are the logarithms of the corresponding components in **d**. With these notations accepted the problem formulation becomes: maximize $h(\mathbf{d}) = -(\ln(\mathbf{d}))^T \mathbf{d}$ under the constraints $f(\mathbf{d}) = \mathbf{j}^T \mathbf{d} - 1 = 0$

and $g(\mathbf{d}) = \mathbf{e}^T\mathbf{d} - E = 0$. This is an example of what is called a constrained optimization problem. The natural symmetry of the problem would be destroyed if we would try to convert the problem into an unconstrained optimization problem by expressing two of the components of $\mathbf{d}$ in the other $s-2$ components. A useful technique that is known as the method of lagrange multipliers, converts the problem into an unconstrained one without disturbing the symmetry. We will use that method for finding the stationary points of $h(\mathbf{d})$.

Suppose $\mathbf{d}'$ satisfies the constraints and $h(\mathbf{d}') \geq h(\mathbf{d})$ for all $\mathbf{d}$ that satisfy the constraints, then

$$\sum_{i=1}^{s} x_i \left[\frac{\partial h}{\partial d_i}\right]_{d=d'} = 0$$

for all $\mathbf{x}$ that satisfy

$$\sum_{i=1}^{s} x_i \left[\frac{\partial f}{\partial d_i}\right]_{d=d'} = 0 \quad \text{and} \quad \sum_{i=1}^{s} x_i \left[\frac{\partial g}{\partial d_i}\right]_{d=d'} = 0.$$

(We assume that there are at least two states with distinct scores.) This implies that there exist a μ and λ such that

$$\forall_{1\leq i\leq s} \left[\left[\frac{\partial h}{\partial d_i}\right]_{d=d'} + \mu \left[\frac{\partial f}{\partial d_i}\right]_{d=d'} + \lambda \left[\frac{\partial g}{\partial d_i}\right]_{d=d'} = 0\right]$$

These s equations together with the two constraints uniquely specify μ, λ and the components of $\mathbf{d}'$. From the s equations

$$-(1 + \ln d_i) + \mu + \lambda e_i = 0$$

it follows that

$$d_i = \exp(\mu + \lambda e_i - 1).$$

And from the two constraints we obtain

$$\sum_{i=1}^{s} \exp(\mu - 1 + \lambda e_i) = \exp(\mu - 1) \sum_{i=1}^{s} \exp(\lambda e_i) = 1$$

or equivalently

$$\sum_{i=1}^{s} \exp(\lambda e_i) = \exp(1 - \mu)$$

and

$$\sum_{i=1}^{s} e_i \exp(\mu + \lambda e_i - 1) = \exp(\mu - 1) \sum_{i=1}^{s} e_i \exp(\lambda e_i) = \frac{\sum_{i=1}^{s} e_i \exp(\lambda e_i)}{\sum_{i=1}^{s} \exp(\lambda e_i)} = E.$$

So, H has a stationary point for

$$\delta(s,\mathrm{t}) = \frac{\exp(\lambda\varepsilon(s))}{\sum_{s'\in\mathcal{S}} \exp(\lambda\varepsilon(s'))}$$

where λ must be such that

$$\frac{\sum_{s\in\mathcal{S}} \varepsilon(s)\exp(\lambda\varepsilon(s))}{\sum_{s'\in\mathcal{S}} \exp(\lambda\varepsilon(s'))} = \mathrm{E}.$$

All second derivatives of the constraint equations, as well as the second derivatives of $\mathrm{h}(\mathbf{d})$ with respect to two different variables are zero. The other second derivatives

$$\frac{\partial^2 \mathrm{h}}{\partial \mathrm{d}_i^2} = -\frac{1}{\mathrm{d}_i},$$

and therefore strictly negative. This shows that h has a local maximum in this point. Since there is only one stationary point this has to be the global maximum.

Theorem 5.3 The equilibrium density of a reflexive chain with an average score E has maximal accessibility if it has the form

$$\delta(s,\mathrm{t}) = \frac{\exp(\lambda(\mathrm{t})\varepsilon(s))}{\sum_{s'\in\mathcal{S}} \exp(\lambda(\mathrm{t})\varepsilon(s'))} \tag{5.6}$$

where λ must be such that

$$\frac{\sum_{s\in\mathcal{S}} \varepsilon(s)\exp(\lambda(\mathrm{t})\varepsilon(s))}{\sum_{s'\in\mathcal{S}} \exp(\lambda(\mathrm{t})\varepsilon(s'))} = \mathrm{E}. \tag{5.7}$$

■

The requirement that λ has to satisfy is not in an explicit form, and consequently it is difficult to obtain a suitable λ from it. But the fact that we want to see predominantly low scores when t gets close to 0, allows us to be a little bit more specific. We therefore split the summation in the denominator into two parts, one adding the terms associated with states with the minimum score ε_0, and one with the other terms. With $\mathcal{S}_0 := \{s|\varepsilon(s) = \varepsilon_0\}$ this becomes

$$\delta(s,\mathrm{t}) = \frac{\exp(\lambda(\mathrm{t})\varepsilon(s))}{\sum_{s\in\mathcal{S}_0} \exp(\lambda(\mathrm{t})\varepsilon_0) + \sum_{s\notin\mathcal{S}_0} \exp(\lambda(\mathrm{t})\varepsilon(s))} =$$

$$= \frac{\exp(\lambda(\mathrm{t})\varepsilon(s))}{|\mathcal{S}_0|\exp(\lambda(\mathrm{t})\varepsilon_0) + \sum_{s\notin\mathcal{S}_0} \exp(\lambda(\mathrm{t})\varepsilon(s))} =$$

$$= \frac{\exp(\lambda(\mathrm{t})(\varepsilon(s)-\varepsilon_0))}{|\mathcal{S}_0| + \sum_{s\notin\mathcal{S}_0} \exp(\lambda(\mathrm{t})(\varepsilon(s)-\varepsilon_0))}$$

Rewriting the equilibrium density in this way makes clear that $\lim_{t\downarrow 0} \lambda(t)$ must be $-\infty$ in order that

$$\delta(s,t) = \begin{cases} \frac{1}{|\mathcal{S}_0|} & \text{if } s \in \mathcal{S}_0 \\ 0 & \text{if } s \notin \mathcal{S}_0 \end{cases}$$

5.3 The acceptance function

From the balance equation it is easy to derive that, given an equilibrium density δ, it has to satisfy

$$\sum_{i=1}^{j-1} (\delta(s_i,t)\alpha(\varepsilon(s_i),\varepsilon(s_j),t)\beta(s_i,s_j) - \delta(s_j,t)\alpha(\varepsilon(s_j),\varepsilon(s_i),t)\beta(s_j,s_i)) =$$

$$= \sum_{i=j+1}^{s} (\delta(s_j,t)\alpha(\varepsilon(s_j),\varepsilon(s_i),t)\beta(s_j,s_i) - \delta(s_i,t)\alpha(\varepsilon(s_i),\varepsilon(s_j),t)\beta(s_i,s_j))$$

for all $1 \leq j \leq s$. (Indices are again with increasing score: $i < j \Rightarrow \varepsilon(s_i) \leq \varepsilon(s_j)$.) These conditions as well as (5.2) and (5.3) would be trivially satisfied with an acceptance function as

$$\alpha(\varepsilon(s),\varepsilon(s'),t) = \min\left\{1, \frac{\delta(s',t)}{\delta(s,t)}\right\} \tag{5.8}$$

if $\forall_{s\in\mathcal{S}}\forall_{s'\in\mathcal{S}}[\beta(s,s') = \beta(s',s)]$. (This condition can be slightly relaxed without diminishing its effect.) It would also make the chain reversible, because even the detailed balance equations are satisfied with such a choice for α. Unfortunately, this would be an extra constraint on a relation very close to the construction of the state space. However, rigorous analysis of annealing chains seems to become very hard, when no extra constraint is imposed on the selection function.

Substituting the result of section 5.2 in (5.8) yields

$$\alpha(\varepsilon,\varepsilon',t) = \min\left\{ 1, \frac{\exp(\lambda(t)\,\varepsilon')}{\exp(\lambda(t)\,\varepsilon)} \right\} = \min\ \{1, \exp\ (\lambda(t)\ (\varepsilon' - \varepsilon))\ \} \tag{5.9}$$

which is identical to the conclusion in theorem 5.2. $\lambda(t)$ has taken the place of $c(t)$, and the constraints on these functions are not contradictory. This acceptance function therefore achieves both, maximal accessibility for a given average score and the trend towards lower scores for lower control values.

If the symmetry of the selection function is accepted, and the acceptance function is chosen conform (5.9), only λ remains to be specified. If S were denumerable infinite, and the scores were uniformly distributed over the positive real axis, it follows from (5.7) that $E(t)\,\lambda(t) = -1$. With $\lambda(t) = -t^{-1}$, a linear behavior of $E(t)$ versus t is then expected. However, ε is bounded below by its minimum value, $\varepsilon_0 > 0$, and mostly also from above. Linear behavior over the whole range of t is therefore impossible with that λ, but in intervals where there are many scores more or less uniformly distributed over that interval a linear relation between t and $E(t)$ should not be surprising.

There are much more important reasons for choosing $\lambda(t) = -t^{-1}$, namely the advantages in controlling and analyzing annealing chains. For example, that choice enforces a simple relation between E and H, for

$$\frac{dE}{dt} - t\frac{dH}{dt} = \sum_{s \in \mathcal{S}} \Big(\varepsilon(s)\dot{\delta}(s,t) + t\dot{\delta}(s,t)\,(1 + \ln(\delta(s,t)))\Big) =$$

$$= \left(t - t\,\ln \sum_{s' \in \mathcal{S}} \exp(-\tfrac{\varepsilon(s')}{t})\right) \sum_{s \in \mathcal{S}} \dot{\delta}(s,t).$$

Putting $\delta(s_i,t) = \frac{d_i}{\sum_j d_j}$, where because of that choice $\dot{d}_k = \frac{\varepsilon(s_k)}{t^2}\,d_k$, makes clear that

$$\sum_{s \in \mathcal{S}} \dot{\delta}(s,t) = \frac{\Sigma_i \Sigma_j \frac{\varepsilon(s_i)}{t^2}\,d_i d_j - \Sigma_i \Sigma_j \frac{\varepsilon(s_j)}{t^2}\,d_i d_j}{\left(\Sigma_j\, d_j\right)^2} = 0,$$

and therefore that $\frac{dE}{dt} = t\frac{dH}{dt}$. Also $\frac{dE}{dt} = \sum_{s \in \mathcal{S}} \varepsilon(s)\dot{\delta}(s,t)$, and using the above notation once more yields

$$\frac{dE}{dt} = \frac{1}{t^2}\left(\frac{\Sigma_i\, \varepsilon^2(s_i) d_i}{\Sigma_k\, d_k} - \frac{\Sigma_i\, \varepsilon(s_i) d_i\, \Sigma_j\, \varepsilon(s_j) d_j}{(\Sigma_k\, d_k)^2}\right) = \frac{\langle \varepsilon^2 \rangle - \langle \varepsilon \rangle^2}{t^2} = \frac{\sigma^2}{t^2}.$$

So, if $\lambda(t) = t^{-1}$, then

$$\frac{dE}{dt} = t\frac{dH}{dt} = \frac{\sigma^2}{t^2}. \tag{5.10}$$

5.4 Properties of annealing chains

- The move set must be such that any state can be reached from any other state via a sequence of moves (*space connectivity*). Further, the move set should be *symmetric* and *reflexive*.

$$\bigcup_{k=1}^{\infty} \mu^k = \mathcal{S} \times \mathcal{S} \tag{3.1}$$

$$\forall_{s \in \mathcal{S}} \forall_{s' \in \mathcal{S}} [(s, s') \in \mu \Rightarrow (s', s) \in \mu]. \tag{3.12}$$

$$\forall_{s \in \mathcal{S}} [(s, s) \in \mu].$$

- The selection function β should be non-zero fif its two arguments are connected by a single move. At least one next state must be selected, so the values that β assigns to all state pairs of which the first one is fixed, must add up to 1. And in this chapter we added the symmetry requirement.

$$\forall_{(s, s') \notin \mu} [\beta(s, s') = 0], \tag{3.2}$$

$$\forall_{(s, s') \in \mu} [\beta(s, s') \neq 0], \tag{3.3}$$

$$\forall_{s \in \mathcal{S}} \left[\sum_{s' \in s\mu} \beta(s, s') = 1 \right], \tag{3.4}$$

$$\forall_{s \in \mathcal{S}} \forall_{s' \in \mathcal{S}} [\beta(s, s') = \beta(s', s)].$$

- The acceptance function is

$$\alpha(\varepsilon(s), \varepsilon(s'), t) = \min \left\{1, \exp\left(t^{-1}\left(\varepsilon(s') - \varepsilon(s)\right)\right)\right\}. \tag{5.11}$$

With a chain specification that satisfies the above requirements the following properties have been ensured:

- The equilibrium density of an annealing chain, with $\varepsilon(s_0) = \varepsilon_0 = \min_{s \in \mathcal{S}} \{\varepsilon(s)\}$, is

$$\delta(s, t) = \frac{\alpha(\varepsilon(s_0), \varepsilon(s), t)}{\sum_{s' \in \mathcal{S}} \alpha(\varepsilon(s_0), \varepsilon(s'), t)} = \frac{\exp\left(-\frac{\varepsilon(s)}{t}\right)}{\sum_{s' \in \mathcal{S}} \exp\left(\frac{-\varepsilon(s')}{t}\right)}. \tag{5.12}$$

and an explicit expression for the score density is therefore

$$\hat{\delta}(\varepsilon, t) = \frac{\hat{\delta}(\varepsilon, \infty)\alpha(\varepsilon_0, \varepsilon, t)}{\Sigma_{\{\varepsilon'\}} \hat{\delta}(\varepsilon', \infty)\alpha(\varepsilon_0, \varepsilon', t)} = \frac{\hat{\delta}(\varepsilon, \infty) \exp(-\frac{\varepsilon}{t})}{\Sigma_{\{\varepsilon'\}} \hat{\delta}(\varepsilon', \infty) \exp(-\frac{\varepsilon'}{t})}. \quad (5.13)$$

- The chain is reversible

$$\forall_{s \in \mathcal{S}} \forall_{s' \in \mathcal{S}} [\delta(s, t)\tau(s, s', t) = \delta(s', t)\tau(s', s, t)]. \quad (3.13)$$

- The eigenvalues of its transition matrix are all real, and, of course, all except one are between -1 and 1. Exactly one eigenvalue is 1. The equilibrium density is independent of β ! Consequently, β can be changed (within the limits imposed on selection functions), without affecting the equilibrium.

- The chain in equilibrium is an ergodic process. This means that in estimating the expectation of the scores only a single realization of the annealing chain has to be observed for a sufficient number of steps.

- For a sufficiently high value of the control parameter t the relative frequency is practically the same for all states.

$$\forall_{s \in \mathcal{S}}[\lim_{t \to \infty} \delta(s, t) = \frac{1}{|\mathcal{S}|}] \quad (5.14)$$

Consequently, the accessibility for high values of t will be close to

$$H_\infty = \ln |\mathcal{S}|. \quad (5.15)$$

- For a sufficiently low value of the control parameter t almost exclusively states with $\varepsilon \approx \varepsilon_0$ will be visited. With $\mathcal{S}_0 = \{s | \varepsilon = \varepsilon_0\}$

$$\lim_{t \downarrow 0} \delta(s, t) = \begin{cases} \frac{1}{|\mathcal{S}_0|} & \text{if } s \in \mathcal{S}_0 \\ 0 & \text{if } s \notin \mathcal{S}_0 \end{cases} \quad (5.16)$$

If s_0 is the only state with $\varepsilon = \varepsilon_0$ (i.e. ε has a unique global optimum over S), the current state will be s_0 for an arbitrarily large proportion of the time for t low enough. If the system has only one global minimum the accessibility will come arbitrarily close to 0 by choosing a t that is low enough. If there are several global minima the accessibility will approach

$$H_0 = \ln |\mathcal{S}_0|. \tag{5.17}$$

- The chain statistics are related according to

$$\frac{dE}{dt} = t\frac{dH}{dt} = \frac{\sigma^2}{t^2}. \tag{5.10}$$

5.5 Discussion

The original justification for using 5.11 as acceptance function was only in the physical analogue, and its implementation in [103]. In this chapter we have tried to give independent reasons for using that function. Other interesting approaches to ensure the convergence to the stationary distribution are in [125], [2] and [128].

In formulating the conditions for the acceptance function that imply the convergence property 5.1 we followed [37], because we want to keep as much freedom as possible in constructing state spaces for optimization problems to which annealing is applied. It also makes clear the price paid for adopting 5.5.

In [128] the more relaxed condition for 5.8 is proven. When $\varepsilon(s) = \varepsilon(s')$ the selection function does not have to satisfy

$$\beta(s, s') = \beta(s', s)$$

as long as

$$\sum_{\{s' | \varepsilon(s') = \varepsilon(s)\}} (\beta(s, s') - \beta(s', s)) = 0$$

is valid.

That all these derivations lead to conditions satisfied by the generally acceptance function makes it reasonable to choose the same function in the remainder of this book. It makes what is going to be said there valid for most current implementations of annealing. In addition, and this is the most important reason, it gives us the validity of 5.10. This relation was already mentioned in [83], but not exploited in the way we will propose in this book.

6 SAMPLES FROM NORMAL DISTRIBUTIONS

The main goal of this chapter is to obtain the result at the end of section 6.3 which is part of the basis for the initialization of annealing as described in section 8.1. On the way to this result we formally define some density functions that are going to be used in chapter 7. After all this effort to obtain theorem 6.14 it is only a small step to the special case of the central limit theorem of section 6.4. This chapter is therefore the background of much that is in chapter 7 and the proof of theorem 8.2. The presentation there is such that it can be read without going through the details of this chapter. These results are achieved by using the characteristic function of a random variable. In section 6.1 this function is defined and some of its more important properties are derived.

6.1 Characteristic functions

Definition 6.1 The *characteristic function* $\varphi(\omega)$ of a random variable x having distribution function F(x) is defined as

$$\varphi(\omega) = \langle e^{j\omega x} \rangle = \int_{-\infty}^{\infty} e^{j\omega x}\, dF(x)$$

where $j = \sqrt{-1}$ and ω is real.

■

Note that if the hth moment exists

$$\frac{d^h\varphi}{d\omega^h} = j^h \int_{-\infty}^{\infty} x^h e^{j\omega x}\, dF(x)$$

so that in particular

Theorem 6.1 The mean and the variance of a random variable x having characteristic function $\varphi(\omega)$ are

$$\mu = \frac{1}{j}\left[\frac{d\varphi}{d\omega}\right]_{\omega=0} \qquad \text{and} \qquad \sigma^2 = \left[\left(\frac{d\varphi}{d\omega}\right)^2 - \frac{d^2\varphi}{d\omega^2}\right]_{\omega=0}$$

if they exist.

■

In sampling theory it is quite often possible to find the characteristic function of a function of the components of a random vector, particularly when that function is linear. The question which arises, of course, is how to find the distribution function of that random variable from its characteristic function. That question is addressed in the following theorem:

Theorem 6.2 (Lévy) Let x be a random variable having characteristic function $\varphi(\omega)$ and distribution function F(x). Then, if F(x) is continuous at $x = x' \pm \Delta, \Delta > 0$, we have

$$F(x' + \Delta) - F(x' - \Delta) = \lim_{\Omega\to\infty} \frac{1}{\pi} \int_{-\Omega}^{\Omega} \frac{\sin\Delta\omega}{\omega} e^{-j\omega x'} \varphi(\omega)d\omega \qquad (6.1)$$

Furthermore, if $\int_{-\infty}^{\infty} | \varphi(\omega) | d\omega < +\infty$, a density function f(x) exists at $x = x'$ and

$$f(x') = \frac{1}{2\pi} \int_{-\infty}^{+\infty} e^{-j\omega x'} \varphi(\omega)d\omega. \qquad (6.2)$$

■

Proof: The fact that $\left| \frac{\sin \Delta\omega}{\omega} e^{-j\omega(x'-x)} \right|$ is bounded (by Δ), justifies changing the order of integration after substituting the definition of $\varphi(\omega)$ in (6.1), and thus we obtain step by step

$$\lim_{\Omega\to\infty} \frac{1}{\pi} \int_{-\Omega}^{\Omega} \frac{\sin \Delta\omega}{\omega} e^{-j\omega x'} \int_{-\infty}^{\infty} e^{j\omega x} dF(x)\, d\omega =$$

$$= \lim_{\Omega\to\infty} \frac{1}{\pi} \int_{-\infty}^{\infty} \int_{-\Omega}^{\Omega} \frac{\sin \Delta\omega}{\omega} e^{j\omega(x-x')} d\omega\, dF(x) =$$

$$= \lim_{\Omega\to\infty} \frac{2}{\pi} \int_{-\infty}^{\infty} \int_{0}^{\Omega} \frac{\sin \Delta\omega}{\omega} \cos(\omega(x-x'))\, d\omega\, dF(x) =$$

$$= \lim_{\Omega\to\infty} \frac{1}{\pi} \int_{-\infty}^{\infty} \int_{0}^{\Omega} \frac{\sin((x-x'+\Delta)\omega)}{\omega} - \frac{\sin((x-x'-\Delta)\omega)}{\omega} d\omega\, dF(x) =$$

$$= \tfrac{1}{2} \int_{-\infty}^{\infty} \left(\frac{x-x'+\Delta}{|\, x-x'+\Delta\,|} - \frac{x-x'-\Delta}{|\, x-x'-\Delta\,|} \right) dF(x) = \int_{x'-\Delta}^{x'+\Delta} dF(x).$$

where the last steps are obtained by moving the limit inside the first integral and making use of the fact that

$$\lim_{\zeta\to\infty} \int_0^{\zeta} \frac{\sin(\alpha z)}{z} dz = \frac{\alpha}{|\alpha|} \frac{\pi}{2}.$$

Since $F(x)$ is continuous at $x = x' \pm \Delta$, we have established (6.1).

Now, to obtain (6.2) we divide (6.1) by 2Δ:

$$\frac{F(x'+\Delta) - F(x'-\Delta)}{2\Delta} = \frac{1}{2\pi} \int_{-\infty}^{\infty} \frac{\sin \Delta\omega}{\Delta\omega} e^{-j\omega x'} \varphi(\omega) d\omega.$$

Note, that $\left| \frac{\sin \Delta\omega}{\Delta\omega} e^{j\omega x'} \varphi(\omega) \right|$ never exceeds $|\,\varphi(\omega)\,|$. If $F(x)$ has a derivative $f(x')$ at $x = x'$, and if $\int_{-\infty}^{\infty} |\,\varphi(\omega)\,|\, d\omega < \infty$, we therefore have

$$\lim_{\Delta\to 0} \frac{F(x'+\Delta) - F(x'-\Delta)}{2\Delta} = \frac{1}{2\pi} \int_{-\infty}^{\infty} \lim_{\Delta\to 0} \left(\frac{\sin \Delta\omega}{\Delta\omega} \right) e^{-j\omega x'} \varphi(\omega) d\omega.$$

$\lim_{\Delta\to 0} \frac{\sin \Delta\omega}{\Delta\omega} = 1$, and substituting that yields (6.2).

qed

Obviously, two random variables x_1 and x_2 having the same distribution function, have identical characteristic functions. It would be convenient if also the converse would be true, since that would provide a basis by which distribution functions may be identified from their corresponding characteristic functions. The following theorem states this as a fact.

Theorem 6.3 Two random variables x_1 and x_2 have identical distribution functions fif their characteristic functions are identical.

■

Proof: Suppose the random variables have the same characteristic function $\varphi(\omega)$. Then if $F_1(x)$ and $F_2(x)$ are the distribution functions of the two random variables it follows from Lévy's theorem (theorem 6.2) that if $x' \pm \Delta$ is *any* interval such that $F_1(x)$ and $F_2(x)$ are continuous at the points $x' \pm \Delta$, then

$$F_1(x'+\Delta) - F_1(x'-\Delta) = F_2(x'+\Delta) - F_2(x'-\Delta)$$

which, together with the fact that $F_1(x)$ and $F_2(x)$ are distribution functions, and therefore continuous on the right, implies that $F_1(x) \equiv F_2(x)$.

qed

In analogy with the definition of a characteristic function of a single random variable, we may define the characteristic function of random vectors.

Definition 6.2 Let $\mathbf{x}$ be a k-dimensional random vector having distribution function $F(\mathbf{x})$. The *characteristic function* $\varphi(\omega)$ of $\mathbf{x}$ is defined as

$$\varphi(\omega) = \langle \exp\left(j\omega^T\mathbf{x}\right)\rangle = \int_{\mathbb{R}^k} \exp\left(j\omega^T\mathbf{x}\right)\, dF(\mathbf{x}).$$

■

It should be observed that the characteristic function of the random vector $(x_1,\ldots,x_h)^T, h < k$, is $\varphi\left((\omega_1,\ldots,\omega_h,0,\ldots,0)^T\right)$ in that case.

By arguments very similar to the one-dimensional case, we can obtain an extension of Lévy's theorem to the case of a random vector with length k.

Theorem 6.4 Let x be a k-dimensional random vector with characteristic function $\varphi(\omega)$ and distribution function $F(x)$. Let I^k be the box $x'_i - \Delta_i < x_i \le x'_i + \Delta_i$, $\Delta_i > 0$, $1 \le i \le k$, in $\mathbb{R}^k_+$. Let $F(x)$ be continuous on the boundary of the box. Then the probability that $x \in I^k$ is

$$\lim_{\Omega \to \infty} \frac{1}{\pi^k} \int_{-\Omega}^{\Omega} \cdots \int_{-\Omega}^{\Omega} \prod_{i=1}^{k} \left(\frac{\sin \Delta_i \omega_i}{\omega_i} e^{-j\omega_i x_i} \right) \varphi(\omega) d\omega_1 \ldots d\omega_k. \tag{6.3}$$

Furthermore, if $\int_{\mathbb{R}^k_+} |\varphi(\omega)| d\omega_1 \ldots d\omega_k < +\infty$, a density function $f(x)$ exists at x' and

$$f(x') = (2\pi)^{-k} \int_{-\infty}^{\infty} \cdots \int_{-\infty}^{\infty} \exp\left(-j\omega^T x'\right) \varphi(\omega)\, d\omega_1 \ldots d\omega_k. \tag{6.4}$$

■

One of the main reasons for introducing characteristic functions in this book is that they are useful in determining whether two random variables are independent. The following theorem is essential for such an application of characteristic functions.

Theorem 6.5 The components of a two-dimensional random vector x, x_1 and x_2, are statistically independent fif $\varphi(\omega_1, \omega_2) = \varphi(\omega_1, 0) . \varphi(0, \omega_2)$ where $\varphi(\omega_1, \omega_2)$ is the characteristic function of x.

■

Proof: Proving that the condition is necessary is trivial. The sufficiency is proved by applying theorem 6.4 for the two-dimensional case. Assume that $\varphi(\omega_1, \omega_2) = \varphi_1(\omega_1)\varphi_2(\omega_2)$, then

$$P(x'_1 - \Delta_1 < x_1 \le x'_1 + \Delta_1, x'_2 - \Delta_2 < x_2 \le x'_2 + \Delta_2) =$$

$$= \lim_{\Omega \to \infty} \frac{1}{\pi^2} \int_{-\Omega}^{\Omega} \int_{-\Omega}^{\Omega} \prod_{i=1}^{2} \left(\frac{\sin \Delta_i \omega_i}{\omega_i} e^{-j\omega_i x'_i} \right) \varphi_1(\omega_1)\varphi_2(\omega_2) d\omega_1\, d\omega_2 =$$

$$= \lim_{\Omega \to \infty} \frac{1}{\pi^2} \left(\int_{-\Omega}^{\Omega} \frac{\sin \Delta_1 \omega_1}{\omega_1} e^{-j\omega_1 x'_1} \varphi_1(\omega_1) d\omega_1 \right) \times$$

$$\times \left(\int_{-\Omega}^{\Omega} \frac{\sin \Delta_2 \omega_2}{\omega_2} e^{-j\omega_2 x'_2} \varphi_2(\omega_2) d\omega_2 \right)$$

which shows that $P(x'_1 - \Delta_1 < x_1 \le x'_1 + \Delta_1, x'_2 - \Delta_2 < x_2 \le x'_2 + \Delta_2) =$

$$P(x'_1 - \Delta_1 < x_1 < x'_1 + \Delta_1) P(x'_2 - \Delta_2 < x_2 < x'_2 + \Delta_2)$$

qed

To conclude this section we mention an easy to prove, and not surprising theorem on the characteristic function of a linear combination of independent random variables.

Theorem 6.6 If $\mathbf{x}$ is a k-dimensional random vector with mutually independent components, and $\varphi_i(\omega_i)$ is the characteristic function of x_i, $1 \leq i \leq k$, then

$$\varphi(\omega) = \prod_{i=1}^{k} \varphi_i(c_i\omega).$$

as the characteristic function of $\mathbf{c}^T\mathbf{x}$.

■

6.2 Quadratic forms and characteristic functions

Theorem 6.7 The function, defined over the entire $\mathbb{R}^k$ as

$$f(\mathbf{x}) = \frac{\exp\left(-\frac{1}{2}(\mathbf{x}-\boldsymbol{\mu})^T\mathbf{S}^{-1}(\mathbf{x}-\boldsymbol{\mu})\right)}{\sqrt{2^k\pi^k\det(\mathbf{S})}}$$

where $\mathbf{S}$ is a positive definite real symmetric matrix, is a density function. The components of the vector $\boldsymbol{\mu}$ are the means of the components of the random vector $\mathbf{x}$. The matrix $\mathbf{S}$ is the covariance matrix of the random vector $\mathbf{x}$.

■

Proof: The verification that the given function is a density function is the same as showing that

$$\int_{\mathbb{R}^k} \exp\left(-\frac{1}{2}(\mathbf{x}-\boldsymbol{\mu})^T\mathbf{S}^{-1}(\mathbf{x}-\boldsymbol{\mu})\right) dy_1\ldots dy_k = \sqrt{2^k\pi^k\ \det(\mathbf{S})},$$

but this is the very contents of theorem 2.23. To prove the other statements of the theorem we introduce another k-dimensional vector variable $\mathbf{q}_k$, and consider the function

$$g(\mathbf{q}) = \int_{\mathbb{R}^k} \exp\left(\mathbf{q}^T(\mathbf{x}-\boldsymbol{\mu})\right) f(\mathbf{x})\, dx_1\ldots dx_k.$$

Let $\mathbf{y} = \mathbf{x} - \mu$, and use a general form of the well-known trick of 'completing the square', i.e.

$$\mathbf{y}^T\mathbf{S}^{-1}\mathbf{y} - 2\mathbf{q}^T\mathbf{y} = (\mathbf{y} - \mathbf{S}\mathbf{q})^T\mathbf{S}^{-1}(\mathbf{y} - \mathbf{S}\mathbf{q}) - \mathbf{q}^T\mathbf{S}\mathbf{q}$$

This enables us to write

$$g(\mathbf{q}) = \frac{1}{\sqrt{2^k\pi^k \det(\mathbf{S})}} \int_{\mathbb{R}^k} \exp\left(\mathbf{q}^T\mathbf{y}\right) \exp\left(-\tfrac{1}{2}\mathbf{y}^T\mathbf{S}^{-1}\mathbf{y}\right) dy_1 \dots dy_k =$$

$$= \frac{\exp\left(\frac{1}{2}\,\mathbf{q}^T\mathbf{S}\mathbf{q}\right)}{\sqrt{2^k\pi^k \det(\mathbf{S})}} \int_{\mathbb{R}^k} \exp\left(-\tfrac{1}{2}(\mathbf{y} - \mathbf{S}\mathbf{q})^T\mathbf{S}^{-1}(\mathbf{y} - \mathbf{S}\mathbf{q})\right) dy_1 \dots dy_k =$$

$$= \exp\left(\tfrac{1}{2}\mathbf{q}^T\mathbf{S}\mathbf{q}\right)$$

In particular, we have

$$\left[\frac{\partial g}{\partial q_i}\right]_{\mathbf{q}=0} = \int_{\mathbb{R}^k} (x_i - \mu_i) f(\mathbf{x}) dx_1 \dots dx_k = \langle x_i - \mu_i \rangle = 0$$

and

$$\left[\frac{\partial^2 g}{\partial q_i \partial q_j}\right]_{\mathbf{q}=0} = \int_{\mathbb{R}^k} (x_i - \mu_i)(x_j - \mu_j) f(\mathbf{x}) dx_1 \dots dx_k = \langle (x_i - \mu_i)(x_j - \mu_j) \rangle = s_{ij}$$

qed

Definition 6.3 The distribution having

$$f(\mathbf{x}) = \frac{\exp\left(-\frac{1}{2}(\mathbf{x} - \mu)^T\mathbf{S}^{-1}(\mathbf{x} - \mu)\right)}{\sqrt{2^k\pi^k \det(\mathbf{S})}}$$

as a density function, with μ being a k-dimensional real vector with k components and $\mathbf{S}_{k,k}$ a positive definite real symmetric matrix, is called the *k-variate normal distribution* and will be denoted by $N(\mu, \mathbf{S})$.

■

Theorem 6.8 The characteristic function of a k-variate normal distribution $N(\mu, \mathbf{S})$ is

$$\varphi(\omega) = \exp\left(-\tfrac{1}{2}\omega^T\mathbf{S}\omega + j\omega^T\mu\right)$$

■

Proof: By definition $\varphi(\omega)$ is equal to

$$\frac{1}{\sqrt{2^k\pi^k \det(\mathbf{S})}} \int_{\mathbb{R}^k} \exp\left(-\tfrac{1}{2}(\mathbf{x} - \mu)^T\mathbf{S}^{-1}(\mathbf{x} - \mu) + j\omega^T\mathbf{x}\right) dx_1 \dots dx_k.$$

$\mathbf{S}$ is positive definite, and thus there is a nonsingular matrix $\mathbf{R}$ such that $\mathbf{S} = \mathbf{R}^T\mathbf{R}$. If we let $\mathbf{z} = (\mathbf{R}^T)^{-1}(\mathbf{x} - \mu)$, the exponent in the integrand reduces to

$$-\tfrac{1}{2}(\mathbf{R}^T\mathbf{z})^T\,(\mathbf{R}^T\mathbf{R})^{-1}\,(\mathbf{R}^T\mathbf{z}) + j\omega^T(\mathbf{R}^T\mathbf{z}) + j\omega^T\mu = -\tfrac{1}{2}(\mathbf{z}^T\mathbf{z}) + j(\mathbf{R}\omega)^T\mathbf{z}.$$

Further, knowing that $\det(\mathbf{S}) = (\det(\mathbf{R}))^2$, we obtain

$$\varphi(\omega) = \frac{\exp(j\omega^T\mu)}{(\sqrt{2\pi})^k}\int_{\mathrm{IR}^k}\exp\left(-\tfrac{1}{2}(\mathbf{z}^T\mathbf{z}) + j(\mathbf{R}\omega)^T\mathbf{z}\right)dz_1\ldots dz_k.$$

Let us concentrate on the integral. With $\mathbf{q} = \mathbf{R}\omega$ that integral is equal to

$$\prod_{i=1}^{k}\left(\int_{-\infty}^{\infty}\exp\left(-\tfrac{1}{2}z_i^2 + jq_iz_i\right)dz_i\right),$$

while

$$\int_{-\infty}^{\infty}\exp\left(-\tfrac{1}{2}z_i^2 + jq_iz_i\right)dz_i =$$

$$= \int_{-\infty}^{\infty}\exp\left(-\tfrac{1}{2}z_i^2\right)\cos(q_iz_i)dz_i + j\int_{-\infty}^{\infty}\exp\left(-\tfrac{1}{2}z_i^2\right)\sin(q_iz_i)dz_i =$$

$$= \int_{-\infty}^{\infty}\exp\left(-\tfrac{1}{2}z_i^2\right)\cos(q_iz_i)dz_i = F(q_i).$$

To evaluate this last integral, we differentiate with respect to q_i and integrate the derivative by parts.

$$\frac{dF}{dq_i} = -\int_{-\infty}^{\infty}\exp\left(-\tfrac{1}{2}z_i^2\right)z_i\,\sin(q_iz_i)dz_i =$$

$$= \left[\exp\left(-\tfrac{1}{2}z_i^2\right)\sin(q_iz_i)\right]_{-\infty}^{\infty} - q_i\int_{-\infty}^{\infty}\exp\left(-\tfrac{1}{2}z_i^2\right)\cos(q_iz_i)dz_i = q_iF(q_i).$$

Solving this little differential equation by separation of variables, and using a previous integral evaluation (theorem 2.23), we obtain

$$F(q_i) = \exp\left(-\tfrac{1}{2}q_i^2\right)\int_{-\infty}^{\infty}\exp\left(-\tfrac{1}{2}z_i^2\right)dz_i = \exp\left(-\tfrac{1}{2}q_i^2\right)\sqrt{2\pi}.$$

Thus, $\int_{\mathrm{IR}^k}\exp\left(-\tfrac{1}{2}\mathbf{z}^T\mathbf{z} + j(\mathbf{R}\omega)^T\mathbf{z}\right)dz_1\ldots dz_k$ equals

$$(\sqrt{2\pi})^k\exp\left(-\tfrac{1}{2}\mathbf{q}^T\mathbf{q}\right) = (\sqrt{2\pi})^k\exp\left(-\tfrac{1}{2}\omega^T\mathbf{S}\omega\right).$$

Substitution of this result in our last expression for $\varphi(\omega)$ completes the proof.

qed

If in this characteristic function we put $\omega_{k_1+1} = \ldots = \omega_k = 0$ we have

$$\varphi(\omega_1,\ldots,\omega_{k_1}, 0,\ldots,0) = \exp\left(j\omega^T\mu - \tfrac{1}{2}\omega^T S^{-1}\omega\right).$$

Hence:

Theorem 6.9 If x is a random vector variable having the k-variate normal distribution $N(\mu, S)$, the marginal distribution of $(x_1,\ldots,x_h)^T$, $h < k$, is the h – variate normal distribution with means $\mu_1,\ldots,\mu_h$ and covariance matrix equal to S with the last $k-h$ rows and columns deleted.

■

The most important distribution function of a continuous random variable is obtained by taking $k_1 = 1$. The distribution is called the *normal* or *Gaussian distribution*. Its density function may be written as

$$f(x) = \frac{1}{\sqrt{2\pi}\sigma}\exp\left(-\frac{(x-\mu)^2}{2\sigma^2}\right)$$

for $-\infty < x < +\infty$, where the parameters μ and σ^2 are the mean and the variance of the distribution. The most convenient form of the normal distribution for tabulation is that corresponding to a random variable y, where $y = (x-\mu)/\sigma$. The density function of y is the *standardized form* N(0,1) of the normal distribution. Note that

$$P(x \le x') = P(y \le y') = \frac{1}{\sqrt{2\pi}}\int_{-\infty}^{y'} e^{-\frac{1}{2}y^2}\,dy$$

where $y' = (x'-\mu)/\sigma$. The distribution function $\Phi(x)$ of the standardized form of the normal distribution defined by

$$\Phi(x) = \frac{1}{\sqrt{2\pi}}\int_{-\infty}^{x} e^{-\frac{1}{2}y^2}\,dy.$$

The problem we consider next is that of finding the distribution of certain quadratic forms of normally distributed random variables. In the general case of k normally distributed random variables having the density function of the k-variate normal distribution, the quadratic form in which we are interested is the one in the exponent of the density function itself, namely, $\frac{1}{2}(x-\mu)^T S^{-1}(x-\mu)$. Consider its characteristic function:

$$\frac{1}{\sqrt{2^k\pi^k\det(S)}}\int_{\mathbb{R}^k}\exp\left(-\tfrac{1}{2}(1-j\omega)(x-\mu)^T S^{-1}(x-\mu)\right)dx_1\ldots dx_k.$$

Substituting according to $\mathbf{y} = (\sqrt{1-j\omega})(\mathbf{x}-\boldsymbol{\mu})$ yields

$$\varphi(\omega) = \frac{1}{\sqrt{2^k \pi^k (1-j\omega)^k \det(\mathbf{S})}} \int_{\mathbb{R}^k} \exp\left(-\tfrac{1}{2}\mathbf{y}^T\mathbf{S}^{-1}\mathbf{y}\right) dy_1 \ldots dy_k.$$

The value of the integral is given in theorem 2.23, and thus

$$\varphi(\omega) = (1-j\omega)^{-\frac{1}{2}k}.$$

The density function of the corresponding distribution is

$$\int_{-\infty}^{\infty} \frac{e^{-j\omega x}\, d\omega}{(1-j\omega)^{\frac{1}{2}k}}$$

An exercise in complex integration will show that the integral vanishes for $x < 0$, and that it can be reduced to Hankel's contour integral for $x \geq 0$. Thus, one may obtain an explicit expression for this density function. However, proving that a certain density function yields the above characteristic function is much easier. Therefore, we prove the following theorem.

Theorem 6.10 If $\mathbf{z}$ is a random vector variable having the 2ν -variate normal distribution $N(\boldsymbol{\mu}, \mathbf{S})$, then $x = (\mathbf{z}-\boldsymbol{\mu})^T\mathbf{S}^{-1}(\mathbf{z}-\boldsymbol{\mu})$ has a density function that is zero for negative x, of course, and for nonnegative x

$$f(x) = \frac{x^{\nu-1}\, e^{-x}}{\int_0^{\infty} u^{\nu-1}\, e^{-u} du} = \frac{x^{\nu-1}\, e^{-x}}{\Gamma(\nu)}$$

This is the density function of the so-called *gamma distribution*, denoted by $G(\nu)$, in which ν is positive, but not necessarily integer. Mean and variance are both equal to ν.

■

Proof: Differentiating $\varphi(\omega) = (\Gamma(\nu))^{-1} \int_0^{\infty} x^{\nu-1} e^{-x} e^{j\omega x} dx$ with respect to ω and integrating by parts yields

$$\frac{d\varphi}{d\omega} = \frac{j}{\Gamma(\nu)} \int_0^{\infty} x^{\nu}\, e^{-(1-j\omega)x}\, dx =$$

$$= \frac{-j}{\Gamma(\nu)(1-j\omega)} \left(\left[x^{\nu} e^{-(1-j\omega)x} \right]_{x=0}^{x=\infty} - \nu \int_0^{\infty} e^{-(1-j\omega)x}\, x^{\nu-1} dx \right) =$$

$$= \frac{j\nu\, \varphi(\omega)}{1-j\omega}.$$

Solving this differential equation, and using $\varphi(0) = 1$ yields

$$\varphi(\omega) = (1 - j\omega)^{-\nu}$$

which is exactly the same function as we came up with in the discussion preceding the statement of the theorem (where $\nu = \frac{1}{2}k$). The uniqueness of characteristic functions guarantees the correctness of the first part of the theorem. Applying theorem 6.1 produces the given values for mean and variance of x.

qed

The integral in the denominator is often used to define the *gamma function* $\Gamma(\nu)$, a function that for positive integer arguments equals the factorial.

Closely related to the gamma distribution is the so-called χ^2 – distribution. this is the distribution of $x = z^T z$, where z is a random vector variable having a k-variate normal distribution $N(\mu, S)$, or in other words the distribution of the sum of squares of normally distributed variables.

Theorem 6.11 If z is a random vector variable having a k - variate normal distribution, then $x = (z - \mu)^T(z - \mu)$ has the density function f(x) which is zero for negative x, and satisfies

$$f(x) = \frac{x^{\frac{1}{2}k-1}\, e^{-\frac{1}{2}x}}{2^{\frac{1}{2}k}\Gamma(\frac{1}{2}k)}$$

for nonnegative x. The distribution of x is then called the *χ^2-distribution with* k *degrees of freedom*, and is denoted by C(k). Its characteristic function is

$$\varphi(\omega) = (1 - 2j\omega)^{-\frac{1}{2}k}$$

its mean and variance are k and 2k, respectively.

■

The proofs are very similar to those of the previous theorem, and can be obtained from them by changing of variables.

6.3 Sampling distributions

Definition 6.4 If x and y are independent random variables that have $F_\eta(x)$ and $F_\zeta(y)$ as distribution functions, and $z = x + y$ has distribution function $F_{\eta+\zeta}(z)$, then $F_\theta(x)$ is said to be a *reproductive* with respect to the parameter θ.

■

Characteristic functions furnish a simple and powerful method of determining sampling distribution functions if the distribution function of the sampled variable is reproductive, because of the following theorem whose straightforward proof is omitted.

Theorem 6.12 A distribution function $F_\theta(x)$ of which $\varphi_\theta(\omega)$ is the associated characteristic function, is reproductive with respect to θ fif

$$\varphi_{\eta+\zeta}(\omega) = \varphi_\eta(\omega)\varphi_\zeta(\omega).$$

■

It is easy to verify that the normal distribution function $N(\mu, \sigma^2)$ is reproductive with respect to both μ and σ^2. Also the gamma distribution function $G(\nu)$ is reproductive with respect to ν.

More in general, suppose $(x_1, \ldots, x_n)$ is a sample from a distribution with distribution function $F_\theta(x)$ and characteristic function $\varphi_\theta(\omega)$. If z is the sample sum then we know from theorem 6.6 that the characteristic function of z is $(\varphi_\theta(\omega))^n$. If now $\varphi_\theta(\omega)$ satisfies the condition of theorem 6.12 then

$$(\varphi_\theta(\omega))^n = \varphi_{n\theta}(\omega).$$

and z has the distribution function $F_{n\theta}(z)$.

Theorem 6.13 If $(x_1, \ldots, x_n)$ is a sample from the distribution with distribution function $F_\theta(x)$, and characteristic function $\varphi_\theta(\omega)$, then $(\varphi_\theta(\omega))^n = \varphi_{n\theta}(\omega)$ implies that the distribution function of the sampling distribution of the sample sum z and sample average $\bar{x}$ are $F_{n\theta}(z)$ and $F_{n\theta}(n\bar{x})$, respectively.

■

This statement can be naturally extended to sampling distributions from k-variate distributions.

Corollary 6.1 If $(x_1, \ldots, x_n)$ is a sample from the k - variate normal distribution $N(\boldsymbol{\mu}, \mathbf{S})$, the sampling distribution of the vector of sample sums $\mathbf{z}$ is $N(n\boldsymbol{\mu}, n\mathbf{S})$. The sampling distribution of the vector of sample averages $\bar{\mathbf{x}}$ is $N(\boldsymbol{\mu}, \frac{1}{n}\mathbf{S})$.

■

Corollary 6.2 If $(x_1, \ldots, x_n)$ is a sample from the gamma distribution $G(\mu)$, the sampling distribution of the of sample sum is the gamma distribution $G(n\mu)$.

■

Corollary 6.3 Let $(x_{11}, \ldots, x_{1n_1}), \ldots, (x_{k1}, \ldots, x_{kn_k})$ be k (independent) samples from $N(\mu_1, \sigma_1^2), \ldots, N(\mu_k, \sigma_k^2)$, respectively. Let $\bar{x}_1, \ldots, \bar{x}_k$ be average of these samples, and let $c_1, \ldots, c_k$ be constants, not all zero. Then $c_1\bar{x}_1 + \ldots + c_k\bar{x}_k$ has as its sampling distribution

$$N\left(\sum_{i=1}^{k} c_i\mu_i, \sum_{i=1}^{k} \frac{c_i^2\sigma_i^2}{n_i} \right).$$

■

The results of this section up to this point are quite easy to obtain and to verify. This is not the case with our next theorem, whose proof will fill the remainder of this section.

Theorem 6.14 If $(x_1, \ldots, x_n)$ is a sample, $\bar{x}$ the sample average, and s^2 the sample variance, then $\bar{x}$ and s^2 are statistically independent fif the samples are taken from a normal distribution $N(\mu, \sigma^2)$. Furthermore, $\sqrt{n}\,(\bar{x} - \mu)/\sigma$, and $(n-1)s^2/\sigma^2$ have as their sampling distributions the normal distribution $N(0,1)$ and the chi-square distribution $C(n-1)$ respectively.

■

Proof: We will use the characteristic function of the joint distribution of $\sqrt{n}(\bar{x}-\mu)/\sigma$ and $(n-1)s^2/\sigma^2$ to prove the independence:

$\varphi(\omega_1, \omega_2) = \langle \exp\left(j\omega_1 \frac{(\bar{x}-\mu)\sqrt{n}}{\sigma} + j\omega_2 \frac{(n-1)s^2}{\sigma^2}\right)\rangle =$

$$= \int_{\mathbb{R}^n} \exp\left(j\omega_1 \frac{(\bar{x}-\mu)\sqrt{n}}{\sigma} + j\omega_2 \frac{(n-1)s^2}{\sigma^2}\right) \prod_{i=1}^{n} \frac{e^{-\frac{(x_i-\mu)^2}{2\sigma^2}}}{\sqrt{2\pi\sigma}} dx_1 \ldots dx_n =$$

$$= (2\pi\sigma)^{-\frac{1}{2}n} \int_{\mathbb{R}^n} \exp -\tfrac{1}{2}(p-q) dx_1 \ldots dx_n,$$

where

$$p = \frac{1}{\sigma_2} \sum_{i=1}^{n} (x_i-\mu)^2 - \frac{2j\omega_2}{\sigma^2} \sum_{i=1}^{n} (x_i - \bar{x})^2 \text{ and } q = 2j\omega_1 \frac{(\bar{x}-\mu)\sqrt{n}}{\sigma}.$$

So if we make, for the moment, the substitutions $\bar{y} = \bar{x} - \mu$, and $\mathbf{y} = \mathbf{x} - \boldsymbol{\mu}$ with $\mathbf{x}$ as an n-dimensional vector of the samples $x_i, 1 \le i \le n$, and $\boldsymbol{\mu}$ as a constant vector, each component equal μ, we obtain for p

$$p = \frac{1}{\sigma^2} \sum_{i=1}^{n} y_i^2 - \frac{2j\omega_2}{\sigma^2} \sum_{i=1}^{n} (y_i - \bar{y})^2 = \frac{1-2j\omega_2}{\sigma^2} \sum_{i=1}^{n} y_i^2 + \frac{4j\omega_2 \bar{y}}{\sigma^2} \sum_{i=1}^{n} y_i - \frac{2j\omega_2 n}{\sigma^2} \bar{y}^2.$$

Since $\bar{y} = \frac{1}{n} \sum_{i=1}^{n} y_i$, this is the same as

$$p = \frac{1-2j\omega_2}{\sigma^2} \sum_{i=1}^{n} y_i^2 + \frac{2j\omega_2}{n\sigma^2} \bar{y}^2.$$

Observing that $\bar{y}^2 = \frac{1}{n^2} \mathbf{y}^T \mathbf{J}_{n,n} \mathbf{y}$, and $\sum_{i=1}^{n} y_i^2 = \mathbf{y}^T \mathbf{I}_{n,n} \mathbf{y}$, we may write

$$p = (\mathbf{x}-\boldsymbol{\mu})^T \mathbf{P}^{-1} (\mathbf{x}-\boldsymbol{\mu}) \text{ with } \mathbf{P}^{-1} = \frac{1-2j\omega_2}{\sigma^2} \mathbf{I}_{n,n} + \frac{2j\omega_2}{n\sigma^2} \mathbf{J}_{n,n}.$$

Also,

$$q = \mathbf{q}^T (\mathbf{x}-\boldsymbol{\mu}) \text{ with } \mathbf{q} = \frac{j\omega_1}{\sigma\sqrt{n}} \mathbf{j}_n.$$

The next step is just another application of 'completing the square':

$$p - q = (\mathbf{x}-\mu)^T \mathbf{P}^{-1}(\mathbf{x}-\mu) - 2\mathbf{q}^T(\mathbf{x}-\mu) =$$

$$= (\mathbf{x}-\mu-\mathbf{Pq})^T \mathbf{P}^{-1}(\mathbf{x}-\mu-\mathbf{Pq}) - \mathbf{q}^T\mathbf{Pq} =$$

$$= (\mathbf{x}-\mu-\frac{j\omega_1}{\sigma\sqrt{n}}\,\mathbf{r})^T \mathbf{P}^{-1}(\mathbf{x}-\mu-\frac{j\omega_1}{\sigma\sqrt{n}}\,\mathbf{r}) + \frac{\omega_1^2}{n\sigma^2}\sum_{i=1}^{n} r_i$$

where $\mathbf{r}$ is the vector of row sums of $\mathbf{P}$. The inverse of matrices like $\mathbf{P}^{-1}$ was given in section 2.1, and therefore

$$\mathbf{P} = \frac{\sigma^2}{1-2j\omega_2}\left(\mathbf{I}_{n,n} - \frac{2j\omega_2}{n}\,\mathbf{J}_{n,n}\right),$$

Thus, $\mathbf{r} = \sigma^2 \mathbf{j}_n$, and we substitute $n\sigma^2$ for $\sum_{i=1}^{n} r_i$ which gives us

$$\varphi(\omega_1,\omega_2) = \exp\left(-\tfrac{1}{2}\omega_1^2\right)\left(\frac{1}{\sigma\sqrt{2\pi}}\right)^n \times$$

$$\times \int_{\mathbb{R}^n} \exp\left(-\tfrac{1}{2}\,(\mathbf{x}-\mu-\frac{j\omega_1}{\sigma\sqrt{n}}\,\mathbf{r})^T \mathbf{P}^{-1}(\mathbf{x}-\mu-\frac{j\omega_1}{\sigma\sqrt{n}}\,\mathbf{r})\right) dx_1 \ldots dx_n.$$

Again, from theorem 2.23 we know that the integral has as its value $\sqrt{2^n\pi^n \det(\mathbf{P})}$, and since

$$\det(\mathbf{P}) = \frac{\sigma^{2n}}{(1-2j\omega_2)^{n-1}},$$

we have

$$\varphi(\omega_1,\omega_2) = \varphi(\omega_1)\cdot\varphi(\omega_2) = \exp(-\tfrac{1}{2}\,\omega_1^2).\ \frac{1}{(\sqrt{1-2j\omega_2})^{n-1}}.$$

That $\varphi(\omega_1,\omega_2)$ factors as required in theorem 6.5 for independence of the corresponding random variables, justifies one part of the theorem. We also have proved the claim about the distributions, because the factors are the characteristic functions of $N(0,1)$ and $C(n-1)$.

We still have to prove that if $\bar{x}$ and s^2 are statistically independent the distribution of the random variables x_i has to be the normal distribution. Let $f(x)$ be the density function of that distribution and

$$\phi(\omega) = \int_{-\infty}^{\infty} e^{j\omega x} f(x)\, dx$$

its characteristic function. Further, let $\varphi_1(\omega_1)$ be the characteristic function of $\bar{x}$ and $\varphi_2(\omega_2)$ be the characteristic function of s^2. The independence or $\bar{x}$ and s^2 implies that $\varphi(\omega_1, \omega_2) = \varphi_1(\omega_1)\varphi_2(\omega_2)$. Using that $\bar{x} = \frac{1}{n}\sum_{i=1}^{n} x_i$ we can express $\varphi_1(\omega_1)$ in terms of ϕ as follows:

$$\varphi_1(\omega_1) = \int_{\mathbb{R}^n} e^{j\omega_1 \bar{x}} f(x_1)\ldots f(x_n)\, dx_1 \ldots dx_n =$$

$$= \int_{-\infty}^{\infty} \cdots \int_{-\infty}^{\infty} \prod_{i=1}^{n} (e^{j\frac{\omega_1}{n}x_i} f(x_i)dx_i) = \left(\int_{-\infty}^{\infty} e^{j\frac{\omega_1}{n}x} f(x)dx\right)^n = \phi^n\left(\frac{\omega_1}{n}\right)$$

Now, we have two ways of expressing $\frac{\partial \varphi}{\partial \omega_2}$ on the line $\omega_2 = 0$. One is using the formal definition of φ, and the other by using

$$\frac{\partial \varphi}{\partial \omega_2} = \varphi_1(\omega_1)\frac{d\varphi_2}{d\omega_2} = \phi^n\left(\frac{\omega_1}{n}\right)\frac{d\varphi_2}{d\omega_2}$$

Of course we still need an expression $\frac{d\varphi_2}{d\omega_2}$ at $\omega_2 = 0$ in the latter method:

$$\left[\frac{d\varphi_2}{d\omega_2}\right]_{\omega_2=0} =$$

$$= \left[\frac{d}{d\omega_2}\int_{\mathbb{R}^n} e^{j\omega_2 s^2} \prod_{i=1}^{n} (f(x_i)dx_i)\right]_{\omega_2=0} = j\int_{\mathbb{R}^n} s^2 \prod_{i=1}^{n} (f(x_i)dx_i) =$$

$$= \frac{j}{n-1}\int_{\mathbb{R}^n} \sum_{k=1}^{n} x_k^2 \prod_{i=1}^{n} (f(x_i)dx_i) - \frac{j}{n(n-1)}\int_{\mathbb{R}^n} \sum_{h=1}^{n}\sum_{k=h}^{n} x_h x_k \prod_{i=1}^{n} (f(x_i)dx_i) =$$

$$= \frac{nj}{n-1}\left(\int_{-\infty}^{\infty} x^2 f(x)dx\right)\left(\int_{-\infty}^{\infty} f(x)dx\right)^{n-1} - \frac{nj}{n-1}\left(\int_{-\infty}^{\infty} xf(x)dx\right)^2\left(\int_{-\infty}^{\infty} f(x)dx\right)^{n-2}$$

$$= \frac{nj}{n-1}\int_{-\infty}^{\infty} x^2 f(x)dx - \frac{nj}{n-1}\left(\int_{-\infty}^{\infty} xf(x)dx\right)^2 = \frac{nj}{n-1}\left(\langle x^2\rangle - \langle x\rangle^2\right) = \frac{nj}{n-1}\sigma^2$$

Having one path to $\frac{\partial \varphi}{\partial \omega_2}$ at $\omega_2 = 0$ completed, we will address the one starting with the formal definition of $\varphi(\omega_1, \omega_2)$ next:

$$\frac{\partial \varphi}{\partial \omega_2} = \frac{\partial}{\partial \omega_2} \int_{\mathbb{R}^n} e^{j\omega_1 \bar{x} + j\omega_2 s^2} \prod_{i=1}^{n} (f(x_i)dx_i) =$$

$$= j \int_{\mathbb{R}^n} s^2 \, e^{j\omega_1 \bar{x} + j\omega_2 s^2} \prod_{i=1}^{n} (f(x_i)dx_i)$$

Thus, for $\omega_2 = 0$ we obtain

$$\left[\frac{\partial \varphi}{\partial \omega_2}\right]_{\omega_2 = 0} = j \int_{\mathbb{R}^n} s^2 \prod_{i=1}^{n} (e^{j\frac{\omega_1}{n} x_i} f(x_i)dx_i) =$$

$$= \frac{nj}{n-1} \phi^{n-1}\left(\frac{\omega_1}{n}\right) \int_{-\infty}^{\infty} x^2 \, e^{j\frac{\omega_1}{n} x} f(x)dx - \frac{nj}{n-1} \phi^{n-2}\left(\frac{\omega_1}{n}\right) \left(\int_{-\infty}^{\infty} x \, e^{j\frac{\omega_1}{n} x} dx\right)^2 =$$

$$= -\frac{nj}{n-1} \phi^{n-1}\left(\frac{\omega_1}{n}\right) \frac{d^2 \phi}{d\left(\frac{\omega_1}{n}\right)^2} + \frac{nj}{n-1} \phi^{n-2}\left(\frac{\omega_1}{n}\right) \left(\frac{d\phi}{d\frac{\omega_1}{n}}\right)^2$$

Equating the two expressions for $\frac{\partial \varphi}{\partial \omega_2}$ at $\omega_2 = 0$, yields the following differential equation

$$\phi^2 \sigma^2 = -\phi \frac{d^2 \phi}{d\omega^2} + \left(\frac{d\phi}{d\omega}\right)^2$$

Substituting $\phi = e^r$ gives the easy equation $\frac{d^2 r}{d\omega^2} = -\sigma^2$. With the constraints $\phi(0) = 1$ and $\left[\frac{d\phi}{d\omega}\right]_{\omega=0} = j\mu$, we find $\phi(\omega) = e^{j\mu\omega - \frac{1}{2}\sigma^2\omega^2}$, which is the characteristic function of the normal distribution.

qed

6.4 Asymptotic properties of sampling distributions

When a sum of samples from a certain distribution has an asymptotically normal distribution, we say that the distribution *satisfies the central limit theorem.* The following theorem states that any distribution with a finite mean and a finite nonzero variance is an example thereof.

Theorem 6.15 If $\bar{x}$ is the average of a sample of size n from a distribution having finite mean μ and finite non-zero variance σ^2, then

$$\lim_{n\to\infty} P\left(\frac{(\bar{x}-\mu)\sqrt{n}}{\sigma} < A\right) = \frac{1}{\sqrt{2\pi}} \int_{-\infty}^{A} e^{-\frac{1}{2}u^2}\, du = \Phi(A).$$

■

Proof: Given is that $\langle x\rangle = \mu < \infty$ and $\langle (x-\mu)^2\rangle = \sigma^2 < \infty$. Let $y = x - \mu$. $F(y)$ and $\varphi(\omega)$ be the distribution function and the characteristic function of y, respectively. If the first and second derivative of φ exist, then we have by Taylor's theorem

$$\varphi(\omega) = \varphi(0) + \omega \left[\frac{d\varphi}{d\omega}\right]_{\omega=0} + \int_0^{\omega} (\omega-\tau) \left[\frac{d^2\varphi}{d\omega^2}\right]_{\omega=\tau} d\tau =$$

$$= 1 + \omega \left[\frac{d\varphi}{d\omega^2}\right]_{\omega=0} + \left[\frac{d^2\varphi}{d\omega^2}\right]_{\omega=0} \int_0^{\omega} (\omega-\tau) d\tau +$$

$$+ \int_0^{\omega} (\omega-\tau) \left(\left[\frac{d^2\varphi}{d\omega^2}\right]_{\omega=\tau} - \left[\frac{d^2\varphi}{d\omega^2}\right]_{\omega=0} \right) d\tau.$$

Since $\langle y\rangle = 0$ we have $\left[\frac{d\varphi}{d\omega}\right]_{\omega=0} = 0$. Also, since $\langle y^2\rangle = \sigma^2$, we know that $\int_{-\infty}^{\infty} y^2\, dF(y) < \infty$, and therefore $\frac{d^2\varphi}{d\omega^2} = -\int_{-\infty}^{\infty} e^{j\omega y} y^2\, dF(y)$ is continuous around $\omega = 0$, which means that there exists a function $\delta(\omega)$ defined as

$$\delta(\omega) = \limsup_{0\le|\tau|\le|\omega|} \left| \left[\frac{d^2\varphi}{d\omega^2}\right]_{\omega=\tau} - \left[\frac{d^2\varphi}{d\omega^2}\right]_{\omega=0} \right|$$

with the property that $\lim_{\omega\to 0} \delta(\omega) = 0$. Thus,

$$\varphi(\omega) = 1 - \tfrac{1}{2}\sigma^2\omega^2 + \vartheta(\omega)\delta(\omega)\omega^2 \text{ where } |\,\vartheta(\omega)\,| \le \tfrac{1}{2}.$$

The characteristic function of $\frac{x-\mu}{\sigma\sqrt{n}} = \frac{y}{\sigma\sqrt{n}}$ is $\varphi(\frac{\omega}{\sigma\sqrt{n}})$, and if we denote the characteristic function of $z = \sum_{i=1}^{n} \frac{x-\mu}{\sigma\sqrt{n}}$ by $\phi_n(\omega)$, we have that

$$\lim_{n\to\infty} \phi_n(\omega) = \lim_{n\to\infty} \varphi^n\left(\frac{\omega}{\sigma\sqrt{n}}\right) =$$

$$= \lim_{n\to\infty} \varphi^n\left(1 - \frac{\omega^2}{2n} + \vartheta\left(\frac{\omega}{\sigma\sqrt{n}}\right)\delta\left(\frac{\omega}{\sigma\sqrt{n}}\right)\frac{\omega^2}{n\sigma^2}\right)^n = \exp\left(-\tfrac{1}{2}\omega^2\right)$$

which is the characteristic function of $N(0,1)$.

The final step in the proof depends on a number of theorems from integration theory which let one prove that

Theorem 6.16 A sequence of distribution functions $\{F_n(z)\}$ is convergent to a distribution function $F(x)$ fif the sequence of associated characteristic functions $\{\phi_n(\omega)\}$ is convergent for every finite ω to a function $\phi(\omega)$, continuous in some neighborhood of the origin. The function $\phi(\omega)$ is then the characteristic function associated with $F(x)$ and the convergence is uniform in every finite interval on the ω-axis.

■

Therefore, it follows that the distribution function of z converges to the distribution function of $N(0,1)$ as n goes to infinity, and since $z = \sqrt{n}(\bar{x} - \mu)/\sigma$, the proof is complete.

qed

6.5 Discussion

The result that the density function of a random variable cannot be normal when $\bar{x}$ and s^2 are statistically dependent, is originally published in [45]. Together with its converse, which can be found in many books on mathematical statistics, it can be used to characterize the normal distribution. This was the main objective in [97]. The proof here follows the latter publication more or less. The other results in this chapter are more familiar, and part of many more advanced studies in mathematical statistics. Theorem 6.2 is usually referenced through [95]. Theorem 6.15 is due to [96].

7 SCORE DENSITIES

A priori knowledge of the behavior of the aggregates is helpful for controlling the course of the algorithm such that chances for a successful run increase. However this behavior depends on the instance. It is therefore interesting to characterize instances with a number of parameters. These parameters do not have to be known at the start of the algorithm. They can be estimated 'on the fly' as long as these estimates have become accurate enough when they are used for deriving decisions. A study of the generic behavior of the aggregates is therefore useful. This behavior depends on the distribution of scores over the states. This distribution is not known in general, and consequently, we have to rely on assumptions and experiments to perform the kind of analysis necessary for discovering the parameters characterizing an instance. For the annealing applications that we have encountered the observations and conclusions of this chapter seem to be in excellent correspondence. However, applying these in implementations has to be done with care and in combination with generally valid criteria. The next chapter will take this task as its subject.

7.1 The density of states

Suppose that the distribution of the scores over the states can be represented by a density function $f : \mathbb{R}_+ \to [0,1] \subset \mathbb{R}$, called the *density of states.*

$$f(\varepsilon) \approx \frac{1}{\Delta\varepsilon \mid \mathcal{S} \mid} \left| \{ s \mid \varepsilon \leq \varepsilon(s) < \varepsilon + \Delta\varepsilon \} \right|. \tag{7.1}$$

This density should be close to the density of the observed scores for $t = \infty$:

$$f(\varepsilon) \approx \hat{\delta}(\varepsilon, \infty),$$

because all states are equally probable for $t = \infty$ by property (5.14) of annealing chains. If this function were completely known, the equilibrium properties of the associated annealing chain would be known for all t. $E(t)$ can then be derived by using the continuous equivalent of (5.13):

$$E(t) = \frac{\int_0^\infty \varepsilon f(\varepsilon)\exp(-\frac{\varepsilon}{t})d\varepsilon}{\int_0^\infty f(\varepsilon)\exp(-\frac{\varepsilon}{t})d\varepsilon} = -\frac{\mathcal{F}'\left(\frac{1}{t}\right)}{\mathcal{F}\left(\frac{1}{t}\right)} \tag{7.2}$$

or equivalently,

$$\mathcal{F}\left(\frac{1}{t}\right) = k\exp\left(\int \frac{E(t)}{t^2}\,dt\right). \tag{7.3}$$

$\mathcal{F}$, in both formulas, is the laplace transform of $f(\varepsilon)$, and k is a constant such that $\mathcal{F}(0) = 1$. If f is a probability density function concentrated on $[0, \infty)$ (i.e. nonnegative and $\int_0^\infty f(\varepsilon)\,d\varepsilon = 1$), then, obviously, $\mathcal{F}$ has derivatives of all orders which satisfy

$$(-1)^n \frac{d^n\mathcal{F}}{dx^n} \geq 0. \tag{7.4}$$

These conditions can be shown to be sufficient as well, but are not very helpful in determining $f(\varepsilon)$ from observations. The first three, for example, are

$$E \geq 0, \qquad E^2 + \sigma^2 \geq 0, \qquad \text{and} \qquad \frac{d\sigma}{dt} \geq -\frac{E^3 + 3\sigma^2 E}{2\sigma}$$

and are amply satisfied in any annealing model. Conditions derived from the higher derivatives are quite complex, and therefore not useful in constructing the density, but they can still be used to check the validity of a proposal for that function.

The above relations indicate several ways of obtaining knowledge about the equilibrium properties of an annealing chain. One way is to estimate the probability law behind the density of states, and derive the implied relation between E and t. This approach is used in section 7.2. Another possibility is to observe the aggregate functions for a certain class of applications, and to construct a parametrized approximation. This is not likely to be consistent with the conditions stated in this section, but we never have a continuous density of states, and the approximation can still be quite useful in a certain range of the control parameter. This line will be followed in section 7.3. There also is the possibility of starting with a laplace transform $\mathcal{F}$, and if its implications are in accordance with the observation, use the inverse laplace transformation to obtain a continuous approximation for the density of states.

7.2 Weak control

In estimating the density of states for a given annealing chain it certainly is important to consider what a priori is known about the scores. It might be that the scores are not only bounded below, but also bounded above, so that all scores are in finite range $[\varepsilon_0, \varepsilon_m]$. Though in general it will not be possible to determine the sharp bounds, some fair estimates are mostly not too difficult to obtain. In any case, the score is required to be positive. Except for such bounds there generally is not much that can be said about the density of states. In such circumstances an unbiased assumption can hardly be different from assigning equal probability to all scores in the range. Indications that $f(\varepsilon)$ differs from a uniform distribution are the values of aggregate functions such as $\sigma^2(t)$. The most likely density function giving rise to such values is the one that maximizes the entropy of the density under the constraints imposed by these values (4.4). More formally stated, maximize

$$\int_{\varepsilon_0}^{\varepsilon_m} f(x)\ \ln(f(x))dx \tag{7.5}$$

under the constraints

$$\int_{\varepsilon_0}^{\varepsilon_m} \rho_k(x)\, f(x)dx = R_k \quad k = 0, 1, 2, \ldots, K \tag{7.6}$$

where $\rho_0 \equiv 1$ and $R_0 = 1$. Using standard techniques of constrained optimization (method of undetermined multipliers) gives the well-known solution

$$f(x) = \exp\left(-1 - \lambda_0 - \sum_{k=1}^{K} \lambda_k \rho_k(x) \right) \tag{7.7}$$

Now suppose that the first two moments of the density $f(\varepsilon)$ have been observed, i.e.

$$\int_{\varepsilon_0}^{\varepsilon_m} \varepsilon\, f(\varepsilon)d\varepsilon = E_\infty \text{ and } \int_{\varepsilon_0}^{\varepsilon_m} \varepsilon^2\, f(\varepsilon)d\varepsilon = \sigma_\infty^2 + E_\infty^2. \tag{7.8}$$

The values of the λ's should then be obtained by evaluating the definite integrals in the constraints, and solving the resulting equations. This leads, in this case, not to manageable expressions. However, it is not unreasonable to ignore the bounds if $E_\infty \gg \sigma$. The consequences of the result can always be compared with results obtained by observing annealing chains, and then it will be clear whether important constraints were excluded. The integrals in the constraints, but taken over the entire $\mathbb{R}_+$ instead of over $[\varepsilon_0, \varepsilon_m]$, are well-known, and determining λ_0, λ_1, and λ_2 from the three equations shows

$$f(\varepsilon) = \frac{1}{\sqrt{2\pi\sigma_\infty^2}} \exp\left(- \frac{(\varepsilon - E_\infty)^2}{2\sigma_\infty^2} \right). \tag{7.9}$$

This result may be not surprising, but it is important to note that it is obtained after ignoring the bounds on the scores that are known to exist, and only using the fact that there is a finite score average and score variance. The implications of (7.9) for the score density when the control parameter assumes finite values is, after (5.13),

$$\hat{\delta}(\varepsilon, t) = \frac{\hat{\delta}(\varepsilon, \infty)\exp(-\frac{\varepsilon}{t})}{\int_{-\infty}^{\infty} \hat{\delta}(\varepsilon, \infty)\exp(-\frac{\varepsilon}{t})\, d\varepsilon}. \tag{7.10}$$

Substituting (7.9) in $\hat{\delta}(\varepsilon, \infty)\exp(-\frac{\varepsilon}{t})$ yields

$$\frac{1}{\sqrt{2\pi\sigma_\infty^2}} \exp\left(-\frac{(\varepsilon - E_\infty)^2}{2\sigma_\infty^2} - \frac{\varepsilon}{t} \right),$$

which can be rewritten as

$$\frac{1}{\sqrt{2\pi\sigma_\infty^2}} \exp\left(\frac{\sigma_\infty^2}{2t^2} - \frac{E_\infty}{t} \right) \exp\left(-\frac{(\varepsilon - (E_\infty - \frac{\sigma_\infty^2}{t}))^2}{2\sigma_\infty^2} \right).$$

In the fraction of (7.10) only the last exponential is not cancelled, because it is the only part of the expression that depends on ε. Consequently,

$$\hat{\delta}(\varepsilon, t) = \frac{\exp\left(-\frac{(\varepsilon - (E_\infty - \frac{\sigma_\infty^2}{t}))^2}{2\sigma_\infty^2} \right)}{\sqrt{2\sigma_\infty^2} \int_{-\infty}^{\infty} \exp(-x^2) dx} = \frac{1}{\sqrt{2\pi\sigma_\infty^2}} \exp\left(-\frac{(\varepsilon - (E_\infty - \frac{\sigma_\infty^2}{t}))^2}{2\sigma_\infty^2} \right). \tag{7.11}$$

This derivation shows that if the scores are normally distributed over the states, the density of observed scores will remain a normal density with a variance not depending on t. The expectation of the observed scores decreases with t according to

$$E(t) = E_\infty - \frac{\sigma_\infty^2}{t}. \tag{7.12}$$

Clearly, this cannot be valid for all values of t, because E is bounded below by ε_0. It is therefore interesting to observe annealing chains derived from the same instance of an optimization problem at different values of t, in particular the average scores of these chains. A result of such an experiment is shown in figure 7.1 in which measured data points and a curve according to (7.12) are plotted. For two decades an excellent correspondence is maintained, but from a certain value down the measured data points clearly deviate from (7.12), and seem to indicate a linear relationship instead. That value will be represented by T. The behavior for $t < T$ is the subject of the next section. The region of $t > T$ is called a region of weak control, because it is not possible to narrow the spread in the scores by lowering t. The average score decreases hyperbolically though, and while it gets closer to ε_0 it has to reach a point where the spread in scores becomes smaller.

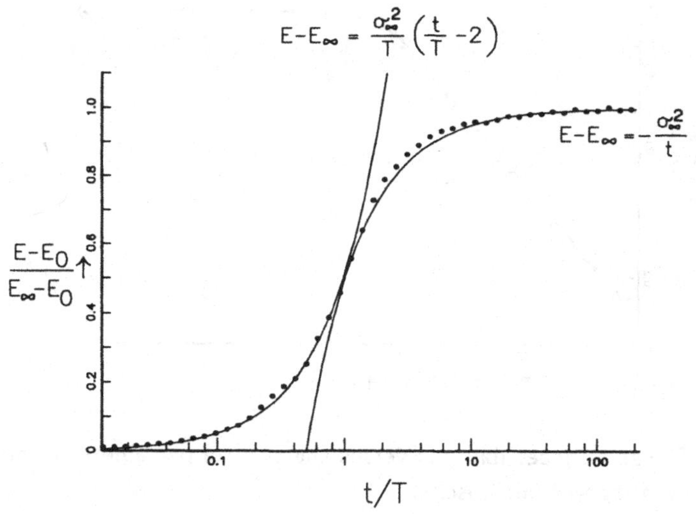

Figure 7.1: The score average E versus the control parameter t, normalized to $\frac{t}{T}$ and on a logarithmic abscissa.

7.3 Strong control

The point $t = T$ is where the deviations from the normal density become noticeable in the plot of E versus t. Surprisingly, the relation between t and E for $t \leq T$ turns out to be mainly linear, down to very low values of t. Because of the relations (5.10) a certain behavior of the score average has an impact on the behavior of the accessibility and the variance. In figure 7.2 the space accessibility is shown as measured for various values of t. This plot clearly confirms what has been observed in the plot of the score average versus t. The full-line curves in that figure 7.2 are not fitted to the measured accessibility, but derived from the linear and hyperbolic relationships discovered in E(t)! The hypothesis of a linear behavior of the score average is therefore strongly supported by the evidence in figure 7.2.

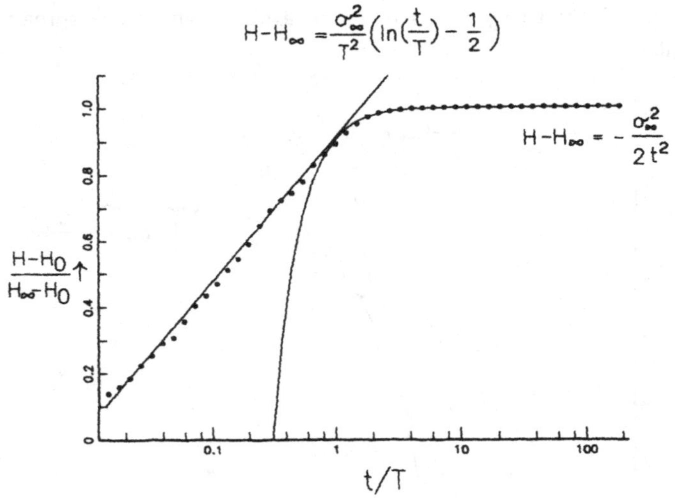

Figure 7.2: The state accessibility H versus the control par ameter t normalized to $\frac{t}{T}$ and on a logarithmic abscissa.

What is the most likely score density causing this linear relationship? To apply the technique of the previous section we have to take into account the fact that the scores are bounded below by some $\underline{\varepsilon} \leq \varepsilon_0$. The observed accessibility does not seem to lead to more manageable expressions. There is, however, another aggregate function, not yet introduced, that is quite helpful in this, namely the expectation of the logarithm of the score

$$L(t) = \sum_{\{\varepsilon\}} \hat{\delta}(\varepsilon, t) \ln(\varepsilon - \underline{\varepsilon}). \tag{7.13}$$

Firstly, because it also exhibits a simple relation with the control parameter, as figure 7.3 shows. There a plot of $L(t)$ versus $\ln(t)$ for the same instance is given, and $L(t)$ seems to be equal to $\ln(t)$, except for an added constant, for $t < T$. Secondly, the constraints (7.6) are easy to evaluate. Note, however, that $L(t)$ can only be obtained after $\underline{\varepsilon}$, the extrapolation of the linear part of $E(t)$ at $t = 0$, is known.

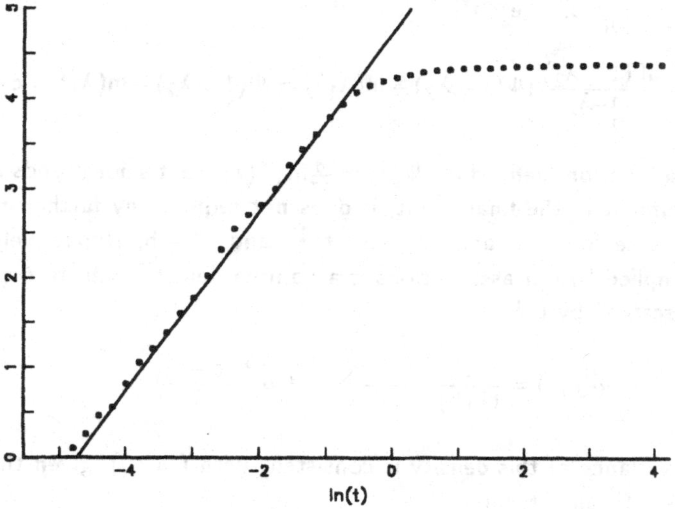

Figure 7.3: The expectation of $\ln(\varepsilon - \underline{\varepsilon})$, $L(t)$, versus $\ln(t)$.

So, we have to obtain the values of the λ's in

$$f(x) = \exp(-1 - \lambda_0 - \lambda_1 x - \lambda_2 \ln x) = a x^{-\lambda_2} \exp(-\lambda_1 x)$$

from the constraints

$$\int_0^\infty \rho_k(x)\, f(x) dx = R_k \quad k = 0, 1, 2 \tag{7.14}$$

with $\rho_0(x) \equiv 1$, $R_0 = 1$, $\rho_1(x) = x$, $R_1 = E(t) - \underline{\varepsilon}$, and $\rho_2(x) = \ln(x)$, $R_2 = L(t)$, where $L(t)$ is the observed average of $\ln(\varepsilon - \underline{\varepsilon})$. By the first constraint the constant a can be expressed in λ_1 and λ_2:

$$\int_0^\infty f(x)dx = a \int_0^\infty x^{-\lambda_2} e^{-\lambda_1 x} dx = \frac{a\Gamma(1-\lambda_2)}{\lambda_1^{1-\lambda_2}} = 1 \;\Rightarrow\; a = \frac{\lambda_1^{1-\lambda_2}}{\Gamma(1-\lambda_2)}$$

assuming that $\lambda_1 > 0$ and $\lambda_2 < 2$. λ_1 and λ_2 have to be determined from the other two constraints and the assumption that $E(t) = bt$ and $L(t) = c + \ln(t)$:

$$\int_0^\infty x f(x) dx = a \int_0^\infty x^{1-\lambda_2} e^{-\lambda_1 x} dx = \frac{a\Gamma(2-\lambda_2)}{\lambda_1^{2-\lambda_2}} = \frac{1}{\lambda_1}\frac{\Gamma(2-\lambda_2)}{\Gamma(1-\lambda_2)} = \frac{1-\lambda_2}{\lambda_1} = bt,$$

$$\int_0^\infty \ln(x) f(x) dx = a \int_0^\infty x^{-\lambda_2} e^{-\lambda_1 x} \ln(x) dx =$$

$$= \frac{a\Gamma(1-\lambda_2)}{\lambda_1^{1-\lambda_2}} (\Psi(1-\lambda_2) - \ln(\lambda_1)) = \Psi(1-\lambda_2) - \ln(\lambda_1) = c + \ln(t).$$

Ψ is Euler's psi function, defined as $\Psi(z) = \frac{d}{dz}\ln(\Gamma(z))$, but since it ends up with a constant argument in the final result, it does not require any further analysis. The obvious choice for λ_1 and λ_2 is t^{-1} and $1-b$, respectively. The score density implied by our assumptions is a gamma density with b degrees of freedom and "scaled" by t^{-1} :

$$\hat{\delta}(\varepsilon, t) = \frac{1}{t\Gamma(b)} \left(\frac{\varepsilon - \underline{\varepsilon}}{t}\right)^{b-1} \exp\left(-\frac{\varepsilon - \underline{\varepsilon}}{t}\right) \tag{7.15}$$

Note that the variance of this density is consistent with (5.10), given the linear relation between E and t, for

$$\sigma^2(t) = -b^2t^2 + \int_{\underline{\varepsilon}}^\infty \frac{(\varepsilon - \underline{\varepsilon})^2}{t\Gamma(b)} \left(\frac{\varepsilon - \underline{\varepsilon}}{t}\right)^{b-1} \exp\left(-\frac{\varepsilon - \underline{\varepsilon}}{t}\right) d\varepsilon =$$

$$= -b^2t^2 + \frac{t^2}{\Gamma(b)} \int_0^\infty y^{b+1} e^{-y} dy = -b^2t^2 + t^2\frac{\Gamma(b+2)}{\Gamma(b)} = bt^2.$$

Adding the variance as a fourth constraint would, therefore, yield $\lambda_3 = 0$, and leave the result unchanged.

7.4 Three parameter aggregates

Accepting the linear dependence of E on t for the major part of $t < T$, and the normal density as a good approximation for $f(\varepsilon)$ when discussing the chains with $t > T$, together with the continuity of the aggregate functions, results in annealing chains of which the aggregate functions are almost completely determined by three parameters, E_∞, σ_∞, and T. Only b in (7.15) has to be expressed in these parameters to obtain expressions for the aggregates in E_∞, σ_∞, and T. Equating the slope of the linear part of E(t) with the first derivative with respect to t of (7.12) at $t = T$ yields

$$b = \frac{\sigma_\infty^2}{T^2}.$$

	$t \geq T$	$t_e < t \leq T$
$E(t) - E_\infty$	$-\frac{\sigma_\infty^2}{t}$	$\frac{\sigma_\infty^2}{T}(\frac{t}{T} - 2)$
$\sigma(t)$	σ_∞	$\sigma_\infty \frac{t}{T}$
$H(t) - H_\infty$	$-\frac{\sigma_\infty^2}{2t^2}$	$\frac{\sigma_\infty^2}{T^2}(\ln(\frac{t}{T}) - \frac{1}{2})$

(7.16)

where t_e is very small, and can be estimated

$$t_e \approx T \exp\left(\frac{T^2(H_0 - H_\infty)}{\sigma_\infty^2} + \frac{1}{2}\right) \tag{7.17}$$

making use of the fact that the accessibility of a chain in equilibrium cannot be below H_0.

When t is decreased to below t_e the second derivative of $E(t)$ will become positive. The exact behavior for $t < t_e$ will, however, strongly depend on the relatively few states with $\varepsilon \approx E(t_e)$ which are very specific to each individual problem.

If the number of degrees of freedom is high the density of (7.15) is hardly distinguishable from the normal density functions. This can be seen by standardizing (7.15) to zero expectation and unit variance, i.e. replacing ε by $xt\sqrt{b} + bt + \underline{\varepsilon}$:

$$\delta^*(x,t) = t\sqrt{b}\delta(xt\sqrt{b} + bt + \underline{\varepsilon}, t) = \frac{b^{b-\frac{1}{2}}e^{-b}}{\Gamma(b)}\exp\left((b-1)\ln(1 + \frac{x}{\sqrt{b}}) - x\sqrt{b}\right)$$

Using Stirling's formula for the Γ-function

$$\Gamma(b) = b^{b-\frac{1}{2}}\, e^{-b}\sqrt{2\pi}\,\left(1 + \mathcal{O}(\tfrac{1}{b})\right)$$

and series expansion for the logarithm

$$\left(\ln(1 + \frac{x}{\sqrt{b}}) = \frac{x}{\sqrt{b}} - \frac{x^2}{2b} + \mathcal{O}((\frac{x}{\sqrt{b}})^3)\right)$$

yields

$$\lim_{b\to\infty} \delta^*(x,t) = \frac{e^{-\frac{x^2}{2}}}{\sqrt{2\pi}}.$$

For a particular instance b is assumed to be constant under the linearity hypothesis. How much the gamma density resembles the normal density for finite b not only depends on the value of b, but also on the value of the scaling factor t^{-1}. In figure 7.4 the densities are shown for several values of t, i.e. gamma densities for $t \leq T$, and normal densities for $t \geq T$. Both curves for $t = T$ are plotted and shown on top of each other. We see that for $t \approx T$ the gamma density is very much like the normal density, while for $t \ll T$ the density is much more asymmetric.

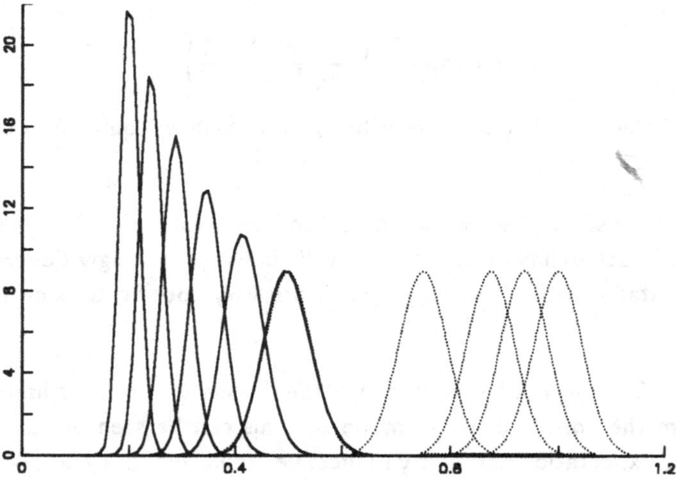

Figure 7.4: The score densities for several values of t. Gamma densities for $t \leq T$ (full) and normal densities for $t \geq T$ (dashed).

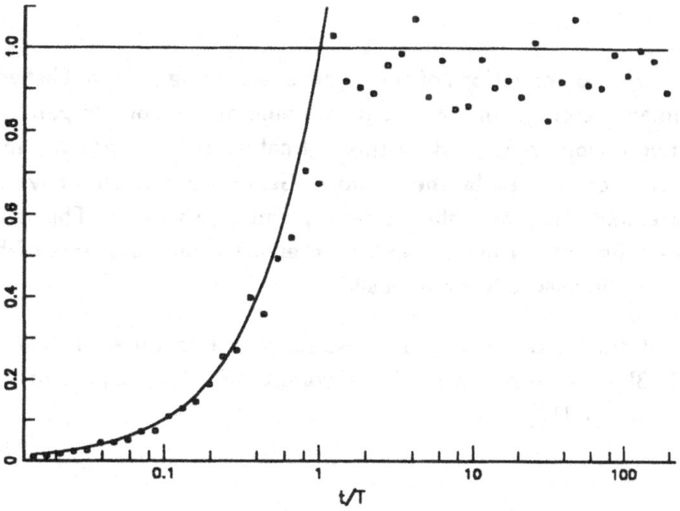

Figure 7.5: The standard deviation of the scores versus the control parameter t (with a logarithmic abscissa).

Although the equations in (7.16) show an excellent correspondence with the observations, they are not consistent with the conditions in section 7.1, since $\mathcal{F}$, derived from (7.16), does not have derivatives of all orders at $t = T$. To eliminate that fact one might replace one of the piecewise expressions in (7.16) by a single analytic formula. It seems expedient, considering the calculus implied by (5.10) and (7.2-7.4), to take a function, expressed by simple relations. (7.4) suggests that the simplest behavior of an aggregate function is to be expected from the standard deviation $\sigma(t)$. Unfortunately, it is also a function of which reliable estimates are difficult to obtain by observing an annealing chain. A rather time consuming experiment produced the plot of figure 7.5. It was observed, however, that the more steps per value of t were allowed, the better the measured data points fitted the expected curve. A good candidate for replacing the piecewise expression for the standard deviation seems to be

$$\sigma(t) = \frac{\sigma_\infty}{2}\left(1 + \nu + \frac{t}{T} - \sqrt{\left(1 + \nu + \frac{t}{T}\right)^2 - 4\frac{t}{T}}\right) \qquad (7.18)$$

It has the nice property that for $\nu \to 0$ it converges to the relations in (7.16). However, even for this expression the calculus needed to obtain $f(\varepsilon)$ is immense.

7.5 Discussion

The observations and derivations of this chapter were originally published in [118]. It also contained the conjecture of a critical value of the control parameter. The presence of two regions separated by this critical value was observed in all applications that were considered by the authors. Generalizing should nevertheless be done with care, and safeguards should be implemented as well. This is usually no problem. The stop criteria in the next chapter are a nice example of how these observations can be used advantageously.

The density of states, often used in physics, was introduced in the annealing context in [153]. The sufficiency of the conditions (7.4) is part of a beautiful theorem by Bernstein[10].

8 THE CONTROL PARAMETER

The goal of the annealing algorithm is to find a state s with $\varepsilon(s)$ close to $\varepsilon(s_0)$. The chain would be almost all the time in such a state after reaching equilibrium for a very low value of the control parameter t , but it would take a huge number of steps to reach and detect that situation. For very high values of t the chain will be almost immediately in equilibrium. Also to reestablish equilibrium from an equilibrium situation at a slightly different t can be done in relatively few steps, because changes in the equilibrium density can be made arbitrarily small by keeping the changes in t small enough. The algorithm therefore performs the annealing chain for several in general decreasing, values of t, each time of course with a limited number of steps, and thereby reducing the space accessibility by confining the chain more and more to states hopefully close to a global minimum. The decrements in t and the number of steps per value of t are the parameters that characterize the *schedule*.

In this chapter the subsequent values of the control parameter are determined. This means that an initial value, the decrements and a final value have to be established. They depend of course on the instance at hand, here characterized by some expected values, to wit the average score, the score variance, and the accessibility. In the first stage the initial values (the values for $t \to +\infty$) of these expectations must be estimated. During the run of the algorithm they are either updated on the basis of new measurements, or estimated anew. Towards the end of the schedule a lot of data is available about the instance, and the observations of the previous chapter concerning the transition point can be brought to bear. As an extra safety the final value of the accessibility, as obtained by updates on the basis of score averages, is compared against the value based on the estimated number of global minima. If these are far apart a new, and slower, annealing may be decided upon. The other important ingredient of the schedule, the length of the individual chains with constant t, is the subject of the next chapter.

8.1 Initialization

Before starting with gradually decreasing t, the algorithm goes through an initialization phase to obtain initial values for t, the accessibility, the score average and variance, and an estimated value for the minimum accessibility, and to set up the data structures that are maintained during the schedule. Among these data structures there is at least one storing the current configuration (state). Preferably, this data structure should uniquely represent one solution to the problem. Having several representations for the same solution makes the state space unnecessarily large, and consequently the algorithm slow. Also, invalid configurations, i.e. configurations that do not correspond to solutions to the real problem, should not be introduced carelessly. They not only enlarge the state space, but might isolate good valid configurations. Another aspect of the state encoding is the ease of implementing the moves and evaluating the scores. These operations have to be performed many times, and should therefore be implemented very efficiently.

H_∞ is known when the size of the state space is known, because $H_\infty = \ln |\mathcal{S}|$ (5.15). $|\mathcal{S}|$, however, is instance dependent, and how to calculate its value depends on the optimization problem. More difficult is it to obtain an estimate for H_0. Its value depends on the number of global minima: $H_0 = \ln \left|\{s \in \mathcal{S} \,|\, \varepsilon(s) = \varepsilon(s_0)\}\right|$. So, if there is only one global minimum, then $H_0 = 0$, but there may be more minima. Quite often the problem has inherent symmetries, so that for every solution a number of equivalent solutions can be constructed. This number can mostly be determined through some combinatorial analysis. In general $\left|\{s \in \mathcal{S} \,|\, \varepsilon(s) = \varepsilon(s_0)\}\right|$ is small in comparison with $|\mathcal{S}|$ and it suffices to estimate that number with the number of combinatorially equivalent solutions.

Estimates for E_∞ and σ_∞ can be obtained by generating enough mutually independent random states $s^{(1)}, s^{(2)}, \ldots, s^{(r)}$ and evaluate ε for these states. It has to be specified however, how many independent states are needed for a desired accuracy in these estimates. Since both, the expectation and the variance of these ε exist, the central limit theorem (theorem 6.15) can be applied for this purpose.

$$P\left(-\frac{\tau\sigma_\infty}{\sqrt{r}} < \frac{1}{r}\sum_{i=1}^{r}(\varepsilon(s^{(i)}) - E_\infty) < \frac{\tau\sigma_\infty}{\sqrt{r}}\right) =$$

$$= \frac{1}{\sqrt{2\pi}}\int_{-\tau}^{\tau} \exp(-\tfrac{1}{2}x^2)dx = \Phi(\tau) - \Phi(-\tau) = 2\Phi(\tau) - 1.$$

Theorem 8.1 The probability that the error in $\frac{1}{r}\sum_{i=1}^{r}\varepsilon(s^{(i)})$ as an estimate for E_∞ is smaller than $\rho\sigma_\infty$ is $2\,\Phi(\rho\sqrt{r}) - 1$.

■

So, it is fairly easy to determine the number of statistically independent states needed to make a reliable estimate for E_∞. It is more difficult to find out how many independent states are needed for keeping the expected relative error in the σ_∞ below a given bound. We know from (4.6) that

$$e = \frac{1}{r-1}\sum_{i=1}^{r}\left(\varepsilon(s_i) - \frac{1}{r}\sum_{j=1}^{r}\varepsilon(s_j)\right)^2$$

is an unbiased estimator for σ_∞^2 : $< e >= \sigma_\infty^2$. The expected relative error in the random variable in e is simply its standard deviation σ_e. To obtain this figure it would be convenient to have its density function available. This function depends on the distribution of the scores. As we saw in chapter 7, they are most likely normally distributed over the states. This allows us to use theorem 6.13, which states that $\frac{(r-1)e}{\sigma_\infty^2}$ is a random variable with density function $C(r-1)$, the χ^2-distribution with $r-1$ degrees of freedom. The characteristic function of $\frac{(r-1)e}{\sigma_\infty^2}$ is according to theorem 6.10

$$(1 - 2j\omega)^{-\frac{r-1}{2}}.$$

Using theorem 6.6 then gives us the characteristic function of the random variable e:

$$\varphi_e(\omega) = \left(1 - 2j\omega\frac{\sigma_\infty^2}{r-1}\right)^{-\frac{r-1}{2}}.$$

The moments of e can now be found by differentiating its characteristic function (theorem 6.1). In particular

$$\sigma_e^2 = \left[\left(\frac{d\varphi_e}{d\omega} \right)^2 - \frac{d^2\varphi_e}{d\omega^2} \right]_{\omega=0} = \frac{2}{r-1}(\sigma_\infty^2)^2.$$

Since the value we want to estimate with e is σ_∞^2, we have the following theorem:

Theorem 8.2 The expected relative error in

$$e = \frac{1}{r-1} \sum_{i=1}^{r} \left(\varepsilon(s_i) - \frac{1}{r} \sum_{j=1}^{r} \varepsilon(s_j) \right)^2$$

as an estimate for σ_∞^2 is

$$\sqrt{\frac{2}{r-1}}. \tag{8.1}$$

■

If the states of $\mathcal{S}$ are all equally likely when the control parameter has its initial value t_b, the score average of the first chain should be equal to the average of scores over the whole state set, that is E_∞. This is not possible, however, because the average score E(t) is lower than E_∞ for any finite, positive t. For a normal distribution of scores the difference between the two averages for high values of t is

$$E_\infty - E(t) = \frac{\sigma_\infty^2}{t}$$

As we observed in chapter 7, the score variance for high t is hardly dependent on t. Choosing t_b much higher than σ_∞ will result in a score density with a slightly lower average and a virtually unchanged variance.

For $t_b = f\,\sigma_\infty,\ f \gg 1$ $\qquad E(t) \approx E_\infty - \frac{\sigma_\infty}{f}$ $\qquad \sigma(t) \approx \sigma_\infty$

To apply the theorems 8.1 and 8.2 it must be possible to generate mutually independent states efficiently. When such a generation procedure is not available, one may set t at a value such that the initial probability of accepting the move with the biggest change in ε is reasonably high. Usually it is not difficult to find an upper bound for the maximum score difference over a move. An initial t then can be calculated for a given probability ξ of accepting such a big change in ε:

$$t_b = \frac{\min_{(s,s') \in \mu}\{\varepsilon(s) - \varepsilon(s')\}}{\ln \xi}.$$

(Minimum because $\ln\xi$ is negative!) After simulating the chain with $t = t_b$ for a sufficient number of steps an estimate for σ^2 is available. If $t_b \gg \bar{\sigma}$ then t_b can be accepted as initial value for the control parameter. During the simulation $\bar{\varepsilon}$ can be updated together with $\bar{\sigma}$. So, if the estimated value of t_b does exceed the estimated standard deviation $\bar{\sigma}$ by a considerable amount, t_b is accepted as the initial control parameter, E_∞ is set to $\bar{\varepsilon}$ and σ_∞ is made equal to the measured standard deviation. If this is not the case, the procedure is to be repeated with a higher value for t_b.

A problem with the latter approach is that the number of steps has to be determined. This number should be high enough to yield reliable values for E and σ^2, and this might be quite high, because the scores observed are not mutually independent. The method using the theorems 8.1 and 8.2 is to be preferred whenever an efficient generation mechanism for mutually independent states is available.

8.2 Decrements in the control parameter

The schedule must be controlled in such a way that the process stays in quasi-equilibrium and yet converges quickly to a global optimum. This has to be achieved by determining the decrements in t, and the number of moves per value for t. The decrements must be chosen such that the steps do not disturb the equilibrium density too much. Of course, we want to generate less states with a score far greater than the current average score E(t), and more states with scores considerably lower than that average. But the relative frequencies of the states closer to the average should not change abruptly at changing t in order to stay close to the equilibrium density $\delta(s,t)$. What is meant by "close to the average" and "close to $\delta(s,t)$" must be specified.

It is clearly not recommendable to interpret "close to average" as having a score differing less than some constant from E(t). If that would be effective for high t, it would inhibit almost any change when t is low. If, on the other hand, it would achieve its goal for low t, the decrements in t will be far too large for keeping the chains in quasi-equilibrium for reasonable chain length (Figure 7.4). It is therefore preferable to measure the difference between a state's score and E(t) in terms of the current standard deviation when determining whether it is "close". For example, states can be considered "close to the average" if their score does not differ more than $\kappa\,\sigma(t)$ from E(t). If we keep the change in relative frequency small for these states when lowering t by an amount of Δt we may expect to stay close to equilibrium (Figure 8.1). This can be stated more formally as

$$\forall_{s\in\{s|E(t)-\kappa\sigma<\varepsilon(s)<E(t)+\kappa\sigma\}}\left[1-\gamma^*<\frac{\delta(s,t-\Delta t)}{\delta(s,t)}<1+\gamma^*\right] \tag{8.2}$$

Another formulation, for small Δt almost the same as (8.2), is

$$\forall_{s\in\{s|E(t)-\kappa\sigma<\varepsilon(s)<E(t)+\kappa\sigma\}}\left[\left|\frac{\dot{\delta}(s,t)}{\delta(s,t)}\right|<\frac{\gamma^*}{\Delta t}\right] \tag{8.3}$$

$\dot{\delta}(s,t)$ being the derivative of $\delta(s,t)$ with respect to t. $\delta(s,t)$ is given by (5.12) and thus

$\dot{\delta}(s,t) =$

$$= \frac{\varepsilon(s)\sum_{s'\in S}\exp\left(\frac{-\varepsilon(s')}{t}\right) - \sum_{s'\in S}\varepsilon(s')\exp\left(\frac{-\varepsilon(s')}{t}\right)}{t^2\exp\left(\frac{\varepsilon(s)}{t}\right)\left(\sum_{s'\in S}\exp\left(\frac{-\varepsilon(s')}{t}\right)\right)^2} =$$

$$= \frac{\varepsilon(s)-E(t)}{t^2}\delta(s,t).$$

More specifically, this yields

Theorem 8.3

$$\forall_{s\in\{s||\varepsilon(s)-E(t)|<\kappa\sigma(t)\}}\left[\left|\dot{\delta}(s,t)\right|<\kappa\frac{\sigma(t)}{t^2}\delta(s,t)\right].$$

■

Using theorem 8.3 in (8.3) gives

$$\forall_{s\in\{s|E(t)-\kappa\sigma<\varepsilon(s)<E(t)+\kappa\sigma\}}\left[\left|\frac{\dot{\delta}(s,t)}{\delta(s,t)}\right|<\kappa\frac{\sigma(t)}{t^2}<\frac{\gamma^*}{\Delta t}\right],$$

because the bound in theorem 8.3 is tight. The requirement (8.3) is, therefore, equivalent to

$$\Delta t = \gamma\frac{t^2}{\sigma(t)}, \tag{8.4}$$

where γ replaces the quotient $\frac{\gamma^*}{\kappa}$.

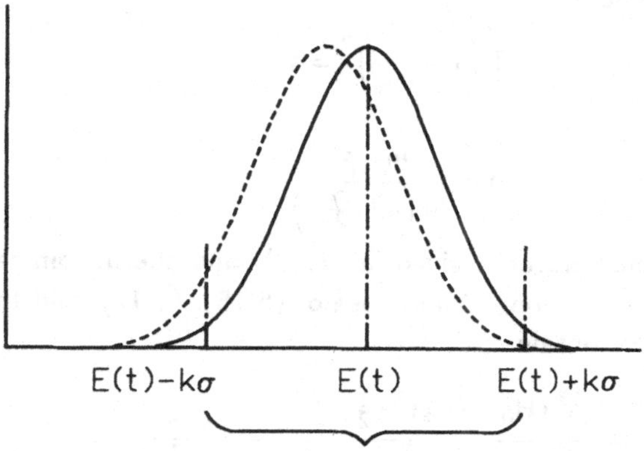

Figure 8.1: The relative frequencies of the states closer to the average should not change abruptly at changing t.

γ should not be chosen independent from the other settings in the algorithm. For example, making $\gamma > \frac{\sigma_\infty}{t_b}$ does not lead to any practical schedule, because the first decrement would be greater than t_b, and the second chain would have a negative, and therefore not permissible, t. It is necessary to require

$$\gamma \, t_b \ll \sigma_\infty. \tag{8.5}$$

Insisting on a combination of γ and t_b violating that requirement can hardly be a well-considered setting, because it would imply playing relatively safe with the initial value of the control parameter, while being careless in preserving quasi-equilibrium. The algorithm should be protected against an ill-suited combination, for example, by preventing $\frac{t-\Delta t}{t}$ to come below a safe value.

With the decrements known, and assuming the generic behavior of the aggregates of chapter 7, it should be possible to get an idea how many chains at different values of t are being performed between the initial value t_b and the estimated value at the end of the inflection region in the generic curve for the score average. If the chains have equal length it tells us something about the computation time required, except for the final stage with $t < t_e$. Most of the time will be spent in the strong control region, that is $t_e < t < T$. t will have passed the value t_e n decrements after passing T, if

$$T\left(1-\frac{\gamma T}{\sigma_\infty}\right)^n = t_e$$

which is the same as

$$n = \frac{\ln\left(\frac{t_e}{T}\right)}{\ln\left(1-\frac{\gamma T}{\sigma_\infty}\right)}.$$

An estimate for the numerator is given in (7.17), and the denominator can be replaced by its series expansion. Making use of (5.15), (5.17), and the fact that $\frac{\gamma T}{\sigma_\infty}$ has to be small, we obtain

$$n = \frac{\frac{T^2}{\sigma_\infty^2}(H_0 - H_\infty) + \frac{1}{2}}{-\frac{\gamma T}{\sigma_\infty} + \mathcal{O}\left((\frac{\gamma T}{\sigma_\infty})^2\right)} \approx \frac{T}{\gamma\sigma_\infty}\ln\left(\frac{|\mathcal{S}|}{|\mathcal{S}_0|}\right).$$

To find the number of decrements in the weak control region we have to solve the difference equation

$$t_{k+1} - t_k + \frac{\gamma}{\sigma_\infty}t_k^2 = 0, \quad t_0 = t_b$$

and determine for which k $t_k < T$. For our application we may replace that difference equation with

$$t_{k+1} - t_k + \frac{\gamma}{\sigma_\infty}t_k t_{k+1} = 0,$$

without risking an unreasonable result. The solution of that equation is

$$\frac{1}{t_k} = \frac{1}{t_b} + k\frac{\gamma}{\sigma_\infty}$$

which gives for the number of decrements while $t > T$

$$k = \left(\frac{1}{T} - \frac{1}{t_b}\right)\frac{\sigma_\infty}{\gamma} \approx \frac{\sigma_\infty}{\gamma T}$$

Combining the two results gives

Theorem 8.4 The total number of decrements as given by

$$\Delta t = \gamma\frac{t^2}{\sigma(t)},$$

if the $\sigma(t)$ is σ_∞ for $t > T$ and $\frac{\sigma_\infty}{T}t$ for $t < T$ is approximately

$$\frac{\sigma_\infty}{\gamma T} + \frac{T}{\gamma\sigma_\infty}\ln\left(\frac{|\mathcal{S}|}{|\mathcal{S}_0|}\right).$$

■

8.3 A stop criterion

If the decrements in t are small enough to keep the process in quasi- equilibrium, a fairly general stop criterion can be used, which can be best understood by studying the behavior of the average score as a function of t. The generic annealing curve of the score average of the chains in equilibrium versus the control parameter t exhibits everywhere a positive derivative and only one inflection region. For lower t the curve flattens towards the minimum. The value of score average where the second derivative becomes positive is already close to the minimum score (Figure 7.1). The maximum improvement that is still possible by decreasing the control parameter from its current value t can be majorized as follows. The maximum improvement is equal to $t\frac{dE}{dt}$ which, with (5.10) can be rewritten as $\frac{\sigma^2}{t}$. This quantity becomes quickly very small compared with the total decrease in E obtained before reaching t (Figure 8.2). This means that

$$\frac{E(t) - E_0}{E_\infty - E(t)} \leq \frac{\sigma^2}{t(E_\infty - E(t))} \tag{8.6}$$

with the right hand side quickly dropping with decreasing t in the region of concavity, i.e. the interval where the second derivative is positive. Although (8.6) does not correctly represent the intended ratio outside that interval, its most likely behavior is staying at 1 for the higher values of t, and then slowly dropping until it enters the interval, in which it shows a much faster drop to zero.

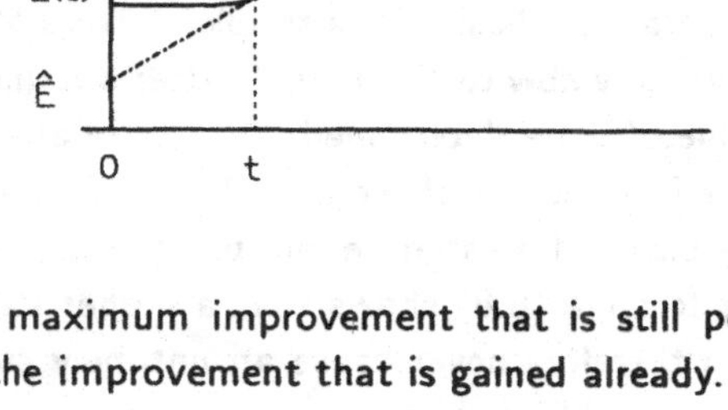

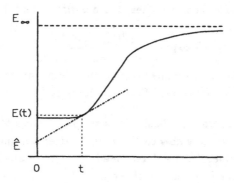

Figure 8.2: The maximum improvement that is still possible should be small compared to the improvement that is gained already.

In practice, however, the process cannot be kept in equilibrium and the aggregates cannot be obtained exactly. Of course, the slower the annealing the better equilibrium is approximated and the more accurate are the aggregates. But we do not want to slow down if not necessary for proper convergence. Applying the test only in the later stages avoids that problem, but requires knowledge about the concavity interval $(0, t_e]$. An elegant solution can be discovered by looking further into E as function of t.

The discussion in chapter 7 indicates that in the intervals $[T, t_b]$ and $[t_e, T]$ the aggregate functions resemble very simple functions of t. These rules are especially simple for the standard deviation of the score. For high t the standard deviation remains fairly constant at its initial value σ_∞. That is where the average score follows the hyperbolic rule. Where the relation between t and the average score is more or less linear, the standard deviation is expected to be proportional to t. We therefore use the standard deviation $\sigma(t)$ for the operational definition of the value of T:

$$T = \sigma_\infty \min_{t>0} \left[\frac{t}{\sigma(t)} \right]. \tag{8.7}$$

In case the behavior is in accordance with the two rules it can be found with the function that gives the dependence of σ on t during this linear decrease: T is the value where evaluation of that function yields σ_∞.

To obtain a safe and useful estimate of the concavity interval we can use (7.17) again, which expresses t_e in accessibilities, score variance and T:

$$t_e = T \exp\left(\tfrac{1}{2} - \frac{T^2 (H_\infty - H_0)}{\sigma_\infty^2} \right). \tag{8.8}$$

σ_∞, H_∞, and H_0 have been determined during the initialization of the algorithm. The only additional estimate we need for using formula (8.8) is the one for T.

The observations so far contain the basics for a simple method of estimating T. In section 8.1 we explained already how to obtain σ_∞. After obtaining σ_∞ the values of σ for each chosen value of t are determined. The estimate for T, T_{est}, each time is the abscissa of the intersection of the line of σ_∞ and the line through the origin and (t, σ) (Figure 8.3). This estimate will be, certainly in the beginning, too high. But analyzing formula (8.8) shows that a higher value of T leads to a lower value t_e, and can therefore never cause an untimely stop. The slope of the line through the origin should be almost constant after coming to values of t

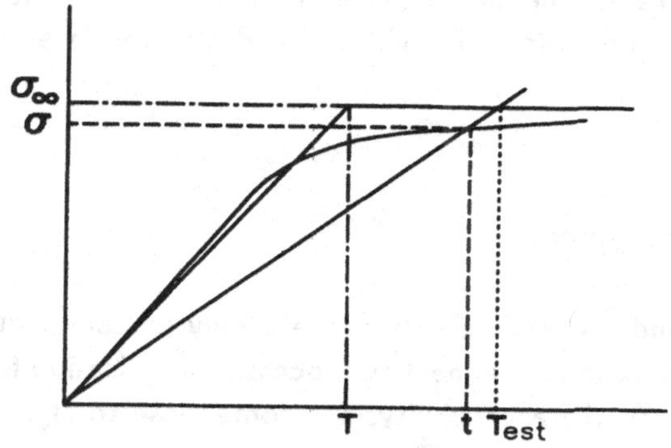

Figure 8.3: Estimating T.

below T, an indication that the process has reached its linear rule. The value of T_{est} stabilizes, and so does the estimate of t_e, calculated with (8.8).

There is one problem with that scheme: How do we obtain reliable values for the score variance? In our experience, using the variance gave the most accurate estimate for T, even though it seems to yield quite wild data for high values t (figure 7.5). However, to achieve this accurateness it was necessary to take precautions against premature conclusions as a consequence of the widely scattered σ values. We therefore apply a smoothing technique. Since we expect the behavior of σ to follow two straight lines double exponential smoothing seems to be the adequate means. The smoothed standard deviation σ_s is calculated with two estimates that are updated in each step: a is an estimate for σ, and b is an estimate for $\frac{d\sigma}{dt}$.

$$\sigma_s = a + b\Delta t \tag{8.9}$$

The updates of a and b are calculated for each newly measured σ in the following way:

$$a_{new} = (1 - \omega_1)\sigma + \omega_1 b_{old}\Delta t \tag{8.10}$$

$$b_{new} = (1 - \omega_2)\frac{a_{new} - a_{old}}{\Delta t} + \omega_2 b_{old} \tag{8.11}$$

ω_1 and ω_2 are real numbers between zero and one that determine the influence of the σ-values of the past. The higher these numbers the more weight is assigned to previous values of the standard deviation, and also the later a change in the slope is discovered.

8.4 Proper convergence

The stop criteria, and also the decrements, are only reliable if quasi-equilibrium has been maintained during the annealing process. An indication for this condition is that the value of H, the accessibility, has come close to H_0. The value of H at high values of t is close to H_∞. The changes in H can be calculated by using (5.10). The value thus obtained for H when the stop criterion is satisfied should be close to H_0. If the schedule is too fast the process is likely to get trapped in a local minimum, for a while or forever. In the latter case H will stay much too high. In the former case H will drop at too low a value for t, and consequently drops quickly and, finally, below H_0 (Figure 8.4). Of course, both effects can occur in the same process, and thereby possibly cancel each others influence. Then, H may be close to H_0 when the stop criterion is satisfied, without convergence to a global minimum.

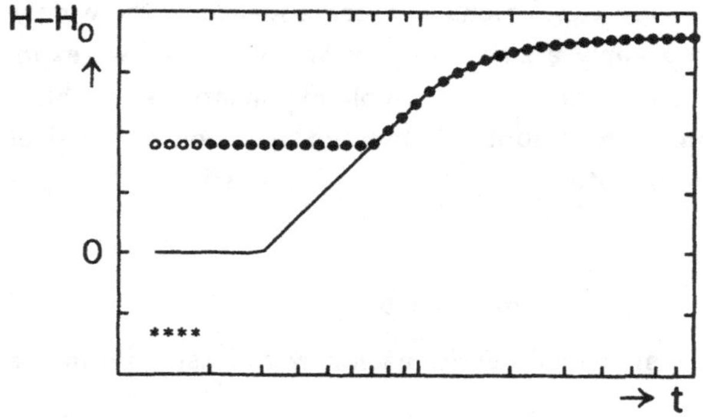

Figure 8.4: The accessibility deviates from H_0 if quasi-equilibrium has not been maintained.

8.5 Discussion

This chapter contains the core of the adaptive strategies in the annealing algorithm. Early results appeared in [119] and [118]. The method of section 8.3 for deriving the stop condition was also presented in [120]. Exponential smoothing was proposed in [19]. The only ingredient missing for a complete adaptive schedule is a way for finding the length of the chains, that is for how long the control parameter should not change after a decrement. This is the subject of section 9.2, which follows [51]. In absence of such an *inner loop criterion* the number of moves for every selected value of t is chosen big enough to obtain useful information about E(t) and σ^2(t). This is then the only requirement when Δt is chosen conservatively.

9 FINITE-TIME BEHAVIOR OF THE ANNEALING ALGORITHM

In chapter 5 we derived that the probability that an annealing chain has s as its current state will go asymptotically to $\delta(s, t)$, the corresponding value of the equilibrium density. Nothing was said however about how fast these probabilities will approach the equilibrium density. Yet it is necessary to know a priori when the actual density is close enough to the equilibrium density to change the value of the control parameter and to start with another chain. In this chapter we want to address that problem.

First we relate the *rate of convergence* of an annealing chain (which is homogeneous markov chain) with the second eigenvalue of the transition matrix. The rate of convergence is interpreted as how fast a measure of deviation from equilibrium is decreasing with the number of steps. This measure of deviation can be defined in many ways and we chose one that leads after a short derivation to a desired relation. There are however many possible definitions that end up with an expression in which the second largest eigenvalue occurs with the number of steps in its exponent.

Next, we derive a lower bound on the number of moves that are necessary before the current state is independent from the initial state. That means that equilibrium cannot be obtained in fewer moves. This lower bound can be interpreted as the ratio between the global and the local accessibility. The local accessibility depends on the transition probabilities , and represents therefore a dynamic aspect of the markov chain. Since the transition probabilities are mostly explicitly known, the local accessibility can be measured at run time. The run time estimates of the speed of convergence can be used for an *inner loop criterion* for the algorithm.

To reach with certainty a global optimum by annealing an infinite time schedule is necessary. This is obviously not practical. Any schedule that has the process pass through more states than the cardinality of the state space is a waste, because a globally optimal configuration can be found by generating all configurations and evaluating their score. A more relevant question is to find the best annealing schedule with limited resources. What is considered best has to be specified of course. In the third section a schedule is considered to be the best when it has the lowest expectation for the score of the final state. The optimization problem discussed there has a high degree of symmetry which makes it possible to derive an analytic expression for that expectation in terms of the control parameter values at the subsequent steps of the annealing algorithm. An optimization routine determines the values that yield the minimum expected final state score.

9.1 Rate of convergence of chains

Running an annealing chain for a number of steps will bring the chain into some state $s \in \mathcal{S}$. The probability that s is the state after n steps depends on n, but will for large n approach $\delta(s, \mathsf{t})$. The rows of $\mathbf{P} = \mathbf{T}^{\mathsf{n}}$ will also approach the vector of state probabilities in equilibrium:

$$\lim_{\mathsf{n}\to\infty} \mathsf{p}_{\mathsf{ij}} = \delta(s_{\mathsf{j}}, \mathsf{t}).$$

This is the contents of the chain limit theorem of section 3. With the *rate of convergence* of the chain we want to indicate how fast these numbers go to their equilibrium values. This means that we have to measure how close the chain is to being in equilibrium. There are many ways to make this intuitive notion more precise, but in the context of annealing it might be interesting to concentrate on the deviations for the states in a particular subset $\mathcal{T} \subset \mathcal{S}$. The next theorem gives an upper bound for one such a *distance* as a function of the number of steps.

Theorem 9.1 For every $\mathcal{T}$, $\emptyset \neq \mathcal{T} \subset \mathcal{S}$, $\mathcal{S}$ being the state set of an annealing chain with transition matrix $\mathbf{T}$,

$$\frac{\rho_-^{\mathsf{n}}}{\min_{s\in\mathcal{T}}\delta(s,\mathsf{t})} \geq \max_{(s_{\mathsf{i}},s_{\mathsf{j}})\in\mathcal{T}\times\mathcal{T}} \left\{ \frac{|\mathsf{p}_{\mathsf{ij}} - \delta(s_{\mathsf{j}},\mathsf{t})|}{\delta(s_{\mathsf{j}},\mathsf{t})} \right\},$$

where $\mathbf{P} = \mathbf{T}^{\mathsf{n}}$ and $\rho_- = \max\{|\lambda| \,|\, \lambda \in \bar{\sigma}(\mathbf{T}) \wedge \lambda \neq 1\}$.

∎

Proof: An annealing chain is reversible, and thus there exists a symmetric matrix $\mathbf{Q} = \mathbf{D}^{\frac{1}{2}}\mathbf{T}\mathbf{D}^{-\frac{1}{2}}$ with the same eigenvalues as $\mathbf{T}$ (all real of course). $\mathbf{D}^{\frac{1}{2}}$ being a diagonal matrix with $\sqrt{\delta(s_h,t)}$ on the diagonal, and $\mathbf{D}^{-\frac{1}{2}}$ as its inverse (theorem 3.9). $\mathbf{Q}$ can also be written as

$$\sum_{h=1}^{s} \lambda_h e^{(h)} e^{(h)T}$$

where $e^{(h)}$ is the eigenvector of $\mathbf{Q}$ associated with λ_h. Let $\lambda_1 = 1$ (there is only one such λ), and let $d_h = \delta(s_h,t)$. Then for every positive n we have

$$\mathbf{P} = \mathbf{T}^n = \mathbf{D}^{-\frac{1}{2}}\mathbf{Q}^n\mathbf{D}^{\frac{1}{2}} = \sum_{h=1}^{s} \lambda_h^n (\mathbf{D}^{-\frac{1}{2}}e^{(h)})(\mathbf{D}^{\frac{1}{2}}e^{(h)})^T =$$

$$= \mathbf{j}\mathbf{d}^T + \sum_{h=2}^{s} \lambda_h^n (\mathbf{D}^{-\frac{1}{2}}e^{(h)})(\mathbf{D}^{\frac{1}{2}}e^{(h)})^T,$$

because $e^{(1)} = \sqrt{\mathbf{d}}$. For individual entries this means

$$p_{ij} = d_j + \sqrt{\frac{d_j}{d_i}} \sum_{h=2}^{s} \lambda_h^n e_i^{(h)} e_j^{(h)}.$$

This yields:

$$\frac{|p_{ij} - d_j|}{d_j} = \frac{\left|\sum_{h=2}^{s} \lambda_h^n e_i^{(h)} e_j^{(h)}\right|}{\sqrt{d_i d_j}} \leq \frac{\rho_-^n}{\min\{d_i,d_j\}} \left| -e_i^{(1)} e_j^{(1)} + \sum_{h=1}^{s} e_i^{(h)} e_j^{(h)} \right| \leq \frac{\rho_-^n}{\min\{d_i,d_j\}}$$

where the orthonormality of the vectors $e^{(i)}$ is used. And thus for any subset $\mathcal{T}$ of $\mathcal{S}$ this implies

$$\forall_{(s_i,s_j)\in\mathcal{T}\times\mathcal{T}} \left[\frac{|p_{ij} - \delta(s_j,t)|}{\delta(s_j,t)} \leq \frac{\rho_-^n}{\min_{s\in\mathcal{T}} \delta(s,t)} \right]$$

qed

This shows that the smaller the modulus of all eigenvalues except unity is, the faster the rate of convergence has to be. Note that $\rho_- = \max\{\lambda_2, -\min(\bar{\sigma}(\mathbf{T}))\}$, where λ_2 is the second largest eigenvalue of $\mathbf{T}$. If no element on the diagonal of $\mathbf{T}$ is less than $\frac{1}{2}$ then $\rho_- = \lambda_2$. This follows directly from the fact that $2\mathbf{T} - \mathbf{I}_s$ is stochastic so that none of its eigenvalues can be smaller than -1 (theorem 3.1). Since $\bar{\sigma}(2\mathbf{T} - \mathbf{I}) = \{2\lambda - 1 | \lambda \in \bar{\sigma}(\mathbf{T})\}$, all eigenvalues of $\mathbf{T}$ must be non-negative. If there are elements less than $\frac{1}{2}$ on the diagonal of $\mathbf{T}$, there may be some negative eigenvalues in the spectrum of $\mathbf{T}$. However, there always exists an annealing chain that is also reversible and has the same equilibrium density δ, and only non-negative eigenvalues. One such a chain has the transition matrix

$$\frac{1}{2(1-a)}\left(\mathbf{T}+(1-2a)\mathbf{I}\right).$$

For this chain $\rho_- \leq \frac{1}{2}(1+\lambda_2)$. To obtain rapid convergence it is therefore important to have a small second eigenvalue. The negative eigenvalues do not play an essential role. It is however not evident how to construct a state space so that the second eigenvalue of the transition matrix is small. In chapter 10 we will return to this subject and link the chain convergence with properties of the state space.

9.2 Minimum number of iterations

Consider a sequence of n successive states $s_0\ s_1 \ldots s_n$ generated by a stationary markov chain. Assume that this markov chain has an equilibrium distribution. Since the density always converges to the equilibrium density, any deviation of the density $f(s_n|s_0)$ must be due to the influence of the density of s_0. Therefore, if s_0 and s_n are independent then $\mathcal{H}(s_n|s_0) = \mathcal{H}(s_n)$.

When the conditional entropy $\mathcal{H}(s_n|s_0)$ is equal to the entropy of the equilibrium density $\mathcal{H}(s_n)$ there is no dependency of the last state s_n on the initial state s_0. This means that equilibrium can always be achieved in n steps irrespective of the initial state. Although we cannot compute this number, it is possible to compute a lower bound for n.

Theorem 9.2 Consider a stationary Markov chain. If s_n and s_0 are statistically independent then

$$n \geq \frac{\mathcal{H}(s_n)}{\mathcal{H}(s_n|s_0)}$$

■

Proof: We will show that $\mathcal{H}(s_n|s_0) \leq n\mathcal{H}(s'|s)$. The entropy of the sequence is larger or equal to the joint entropy of the first and the last state of the sequence:

$$\mathcal{H}(s_0\, s_1 \ldots s_n) \geq \mathcal{H}(s_0\, s_n) \tag{9.1}$$

Using the series expansion of (4.2) on both sides gives:

$$\mathcal{H}(s_0) + \sum_{i=1}^{n} \mathcal{H}(s_i|s_{i-1} \ldots s_0) \geq \mathcal{H}(s_0) + \mathcal{H}(s_n|s_0) \tag{9.2}$$

Eliminating $\mathcal{H}(s_0)$ and using the markov property yields:

$$\sum_{i=1}^{n} \mathcal{H}(s_i|s_{i-1}) \geq \mathcal{H}(s_n|s_0) \tag{9.3}$$

and since the chain is stationary the conditional entropies do not depend on i :

$$n\,\mathcal{H}(s'|s) \geq \mathcal{H}(s_n|s_0) \tag{9.4}$$

If s_n and s_0 are independent then $\mathcal{H}(s_n) = \mathcal{H}(s_n|s_0)$.
Therefore $\mathcal{H}(s_n) \leq n\mathcal{H}(s'|s)$

qed

If the markov chain is an annealing chain the global accessibility $H_t = \mathcal{H}(s_n)$ and the local accessibility $h_t = \mathcal{H}(s'|s)$. The global accessibility H_t may be interpreted as a measure for the size of the likely state space. The local accessibility h_t can be seen as measuring the access to neighboring states. The inner loop criterion compares the local with the global accessibility. Their ratio will be called the *minimum dependence period* $P = H/h$. It may be interpreted as the 'size' of the state space relative to the speed of 'dispersion'. If the change in t is small enough, it is reasonable to assume that the chain is stationary. Equilibrium can only be obtained from non-equilibrium in a period longer than the minimum dependence period.

The minimum dependence period can now be computed, since we also know

$$H_t = H_\infty + \int_\infty^t \frac{dE}{t} \tag{9.5}$$

For many practical systems all states are equally likely when $t = \infty$ and $H_\infty = \ln(|\mathcal{S}|)$.

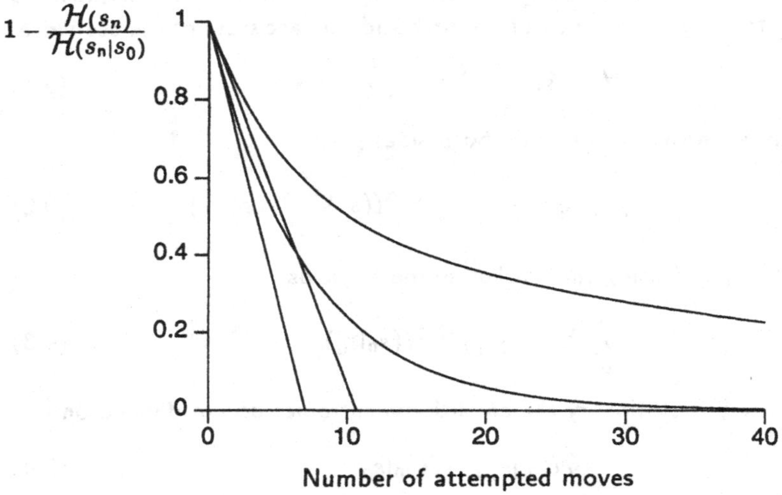

Figure 9.1: The speed of convergence

In figure 9.1 the computation of P is illustrated. The conditional entropy $\mathcal{H}(s_n|s_0)$ approaches the accessibility H as the number of iterations increases. The top curve illustrates an annealing chain at high t, when all states are still easily accessible. The bottom curve illustrates a chain at low t, when it is not easy to go from one valley to another. The tangent lines are the linear extrapolation of local accessability. According to the theorem the curves cannot be higher than their associated tangent lines. The minimum dependence period is the value of n at which the tangential reaches the value H.

A run time estimate of the minimum dependence period P can be made. This is possible because the transition probabilities of most moves are known explicitly. The transition probability of a move is the product of a selection probability and an acceptance probability. Three kinds of moves can be distinguished: upward moves, downward moves and identity moves. Each kind has its own transition probability.

$$\tau(s, s\prime, \mathrm{t}) = \begin{cases} \beta(s, s\prime)\exp(\frac{\varepsilon(s)-\varepsilon(s\prime)}{\mathrm{t}}) & \text{if } \varepsilon(s) < \varepsilon(s\prime) \wedge s \neq s\prime \\ \beta(s, s\prime) & \text{if } \varepsilon(s) \geq \varepsilon(s\prime) \wedge s \neq s\prime \\ \tau(s, s, \mathrm{t}) & \text{if } s = s\prime \end{cases} \tag{9.6}$$

The expectation of the local accessibility can be replaced by the average over a sequence of successive states.

$$h_t = \mathcal{H}(s'|s) = \lim_{k\to\infty} -\frac{1}{k}\sum_{i=1}^{k} \ln(\tau(s_{i-1}, s_i, t)) =$$

$$= -\lim_{k\to\infty}\left[\frac{1}{k}\sum_{\varepsilon(s_i)>\varepsilon(s_{i-1})} \ln(\beta(s_{i-1}, s_i)\exp(\frac{\varepsilon(s_{i-1})-\varepsilon(s_i)}{t}))+\right.$$

$$\left.+\frac{1}{k}\sum_{\varepsilon(s_i)\le\varepsilon(s_{i-1})} \ln(\beta(s_{i-1}, s_i)) + \frac{1}{k}\sum_{s_i=s_{i-1}} \ln(\tau(s_i, s_i, t))\right] =$$

$$= \lim_{k\to\infty} \sum_{s_i\neq s_{i-1}} \frac{-\ln(\beta(s_{i-1}, s_i))}{k} + \tag{9.7}$$

$$+\lim_{k\to\infty}\left[\sum_{\varepsilon(s_i)\ge\varepsilon(s_{i-1})} \frac{\varepsilon(s_i)-\varepsilon(s_{i-1})}{kt} - \sum_{s_i=s_{i-1}} \frac{\ln(\tau(s_i, s_i, t))}{k}\right]$$

By replacing the infinite number of states by a finite number k of actually observed states, an estimate of the local accessibility can be made during run time. By taking a sufficiently large number k it is possible to determine the local accessibility at run time with reasonable accuracy.

A problem is that $\tau(s, s, t)$, the probability of an identity or rejected move, is difficult to determine. Let $a = |\{s_i | s_i \neq s_{i-1}\}|/k$, the fraction of moves that changed the state. a is an estimator for $\langle 1-\tau(s, s, t)\rangle$. Using the fact that ln is a convex function a bound can be derived for the last term of h_t

$$-\frac{1}{k}\sum_{s_i=s_{i-1}} \ln(\tau(s_i, s_i, t)) \le (a-1)\ln(\sum_{s_i=s_{i-1}} \frac{\tau(s_i, s_i, t)}{k(1-a)}) = (a-1)\ln(1-a) \tag{9.8}$$

This bound gets tighter as $a \to 0$. In the case that all non-zero selection probabilities are equal, $\beta(s, s') = \beta$, the local accessibility can be estimated by:

$$h_t \approx a\ln(\frac{1}{\beta}) + \frac{\overline{\max(0, \Delta\varepsilon)}}{t} + (a-1)\ln(1-a) \tag{9.9}$$

This formula shows that the rate of convergence becomes slower when the number of accepted moves decreases. Less intuitive is the fact that a smaller average score change of the upward moves indicates slower convergence.

The minimum dependence period can be used as an inner loop criterion for the number of moves that must be done after a step of t. Common heuristics such as keeping the number of accepted moves equal to the size of the problem can be replaced by a more general criterion, based on theoretical considerations. An automatic schedule control with this criterion will adapt to the roughness of the landscape.

A problem with this method is the lack of precision in determining the accessibility H. The integration errors are accumulated and become relatively large as H approaches 0. It may even happen that the estimate of H becomes negative. Because the convergence cannot improve as t decreases, the maximum value of P encountered so far is retained and used as inner loop criterion.

9.3 Finite-time optimal schedules

The annealing algorithm can be modeled as an inhomogeneous markov chain, that is a process with transition probabilities that change over time. The transition probabilities over several steps of the algorithm are no longer the entries of the power of a (constant) single-step transition matrix when the control parameter is changed over these steps. The general expression for the probability density of the states is a product of a number of matrices and the initial density function:

$$\mathbf{p}^n = \mathbf{T}(t_n)\mathbf{T}(t_{n-1})\cdots\mathbf{T}(t_1)\mathbf{p}^0. \tag{9.10}$$

Here is $\mathbf{p}^i$ the vector of probability densities just before the $(i+1)$-th transition. When the scores of the states and the move set are known, the subsequent transition matrices can be calculated, and the probability of any state to be the final state can be evaluated by using formula 9.10. For all problems of interest this is impractical. Typical annealing problems have huge state spaces, and calculating all these matrices for the usually many steps is a task many times more consumptive than evaluating all scores, and the latter is already impractical. But for problems with a simple structure and a moderate size (and therefore not requiring annealing) we can try to find an explicit expression for $\mathbf{p}^n$. Optimizing the inner product of the density vector $\mathbf{p}^n$ with the vector of scores over all possible settings of $t_0, t_1, \ldots, t_n$ for given n yields the best schedule of length n. That is the sequence of values for the control parameter that minimizes the expected value of the score of the final state. This experiment has been carried out for a number of small size partition problems. They work on the complete graph with 2^L vertices. The edges in this graph have an individual weight. The vertices are to be divided into two blocks of 2^{L-1} vertices such that the sum of the weights of edges connecting a vertex in one block with a vertex in the other block is minimum. The assignment of the weights to the edges is done by establishing a bijective relation between the vertices of the graph and the leaf nodes of a binary tree of depth L. The weight of an edge between two vertices is then made equal to a^k where a is a real number greater than 1 and k is the level of the nearest common ancestor of the two vertices:

$$k = \max\{h \mid \text{the vertices have a common ancestor at level } h\}.$$

The weights of the edges are therefore between 1 and a^{L-1}. The state space for this problem is as follows. The states are the partitions of the vertex set with two equally-sized blocks. The move set contains all interchanges of a vertex from one block with a vertex from the other block. The scores are the sum of the weights of edges connecting vertices from different blocks.

The problem has two global minima, namely the ones where the vertices corresponding with the leafs of the left subtree of the root are in the same block. Local minima occur already when $L = 3$. In that case 6 out of 35 states are local minima. Only 2 of these are global minima, and so there is a chance on getting stuck at a non-global minimum. The number of states grows quickly with L. For $L = 4$ there are already more than 10000 states. Analysis is however simplified by the fact that the states fall into a small number of classes that makes it feasible to treat the problem in a condensed form. For $L = 3$ there are 5 such classes, for $L = 4$ this number is 35. A complete analysis was carried out for $L = 3$ and $a = 3$.

To find the optimal schedule the following objective function has to minimized over the set of all possible $\mathbf{t}^T = (t_1, t_2, \ldots, t_n)^T \in \mathbb{R}^n$:

$$\mathbf{e}^T \mathbf{T}(t_n) \mathbf{T}(t_{n-1}) \cdots \mathbf{T}(t_1) \mathbf{p}^0, \tag{9.11}$$

where $\mathbf{e}$ is the vector of scores. In our definition of annealing we only allowed a non-increasing sequence of positive values for the control parameter. Dropping the positivity constraint had no consequence for the analysis (negative t did not occur in optimal schedules), while dropping the monotonicity yields interesting results. The minimization of this (non-linear) objective function has been carried out with a standard optimization package. The surprising result was that all locally optimal schedules start and end with a number of zeros. That the final values of t have to be 0 can be proven, but that the initial values are zeros is against the prevailing intuition. The number of zeros at the end are different for different schedule lengths, but the number of initial zeros for globally optimal schedules seems to be independent of n. For $n \leq 14$ the globally optimal schedule is the one with all zeros. That means that for these schedule lengths iterative improvement rather than annealing with positive t is to be used. Iterative improvement is always locally optimal. For schedules longer than 14 steps, however, the lowest expected final score is achieved with schedules with some positive values for t. They also hit more often a global minimum than the all-zero (that is iterative improvement) scheme. For all $14 < n \leq 70$ the schedule has first four zeros for the control parameter, then a jump towards a positive value, followed by a monotonic decrease to zero, which is reached 8 to 10 steps before the end. An explanation one can give for the initial zeros is that it makes no sense to apply annealing when there is only a slim chance to get stuck in a local minimum. The initial zeros, therefore, bring it at the level of the first local minima. That for longer schedules the number of zeros at the end decreases is reasonable, because a longer schedule allows a smoother approach to zero than a schedule that only has a few steps available.

9.4 Discussion

The way we measure the deviation from the equilibrium density is rather arbitrary. It was introduced in [139] where also theorem 9.1 is proven. Other definitions have been proposed that also lead to an expression that indicates that convergence is geometrically fast with the second largest eigenvalue of the transition matrix as the common ratio. One such a statement can be found in [138] where is shown that

$$\mathbf{T}^n = \mathbf{T}^\infty + \mathcal{O}(n^{(m_2-1)}|\lambda_2|^k)$$

when $\lambda_2 \neq 0$. The proof is based on the Perron-Frobenius theorem, a theorem we avoided in this book (also in proving this theorem the full Perron-Frobenius theorem is not necessary). Unfortunately, it is not easy to derive guidelines for state space construction from the fact that a small second eigenvalue is beneficial for the convergence of an annealing chain.

The inner loop criterion of section 9.2 was first published in [51]. It is to our knowledge the only criterion found so far that is quite general and at the same time simple to implement. It does however not express what can be achieved by decreasing the control parameter slowly. The smaller the decrease in t when the chain is in equilibrium, the sooner the chain is expected to be in equilibrium again. A relation reflecting that expectation would provide a more solid basis for the trade-off between chain length and control parameter decreases.

The results on optimal finite-time schedules are all due to Philip N. Strenski. The partitioning problem was formulated by Greg B. Sorkin. Although submitted for publication Strenski's paper has not yet appeared as far as we know. In that case it can only be obtained as a research report from Thomas J. Watson Research Center of the IBM Corporation in Yorktown Heights (NY). More interesting details can be found in that report. He has an explanation for the initial zeros by using the physical analogy: A uniform probability density for the state set corresponds with infinite value for the control parameter (when not all scores are equal), and in the physical analogue with an infinite temperature. The internal energy of the system is, however, bounded, and any real, finite-energy system has to have a finite temperature. Since the control parameter corresponds rather with the temperature of an external thermal bath than with the internal temperature of the system, the most efficient way of bringing the system at a temperature that it can support, is to cool as fast as possible, and that is with an external bath at the lowest possible temperature.

10 THE STRUCTURE OF THE STATE SPACE

To formulate an optimization problem for the annealing algorithm we have to specify the state space $(\mathcal{S}, \mu)$, the score function $\varepsilon(s)$, and the selection function β. The states and the score function are strongly suggested by the problem. Often we may choose to optimize an estimate of the real object function, because computing that function would be too time consuming. Also we may restrict the search to a subset of all possible configurations, because we know that this subset contains, if not an optimum configuration, plenty of good solutions. Nevertheless, the options for manipulating the state set and the score function are quite limited. In contrast, there usually is a wide range of possibilities in selecting a move set when implementing the annealing algorithm for a given problem. Of course, the move set has to be transitively closed, reflexive and symmetric, but that still leaves substantial latitude, and therefore the question, what has to be taken into consideration in using that freedom.

In this chapter we first want to confirm our intuition that the structure of the state space has an influence on the convergence properties of the chains. In a sense this is already expressed in theorem 9.1, because the eigenvalues depend on the state space. Here a new quantity, the space conductance, is defined, and a relation between this quantity and the second eigenvalue is derived. Then we develop a terminology for state space properties and translate intuition about the adequacy of a state space into the new terms. This terminology also enables us to phrase a fundamental result on the global convergence of the annealing algorithm. It certainly is unsatisfactory that we cannot give a recipe for state space construction. Not even a useful space analysis method can be given. Only some intuitive considerations about what is desirable in state spaces for annealing have to conclude this chapter about the most important topic for annealing applications: constructing the state space.

10.1 Chain convergence

If there is only a small total probability that a state in $\mathcal{T} \subset \mathcal{S}$ is the current state, and the probability of leaving $\mathcal{T}$ is relatively low, slow convergence should be expected. To characterize the situation we define the *space conductance*:

$$\Phi_{\mathcal{S},\mu}(t) =$$

$$= \min \left\{ \frac{\sum_{(s,s')\in\mathcal{T}\times(\mathcal{S}\setminus\mathcal{T})} \delta(s,t)\tau(s,s',t)}{\sum_{s\in\mathcal{T}} \delta(s,t)} \middle| \emptyset \subset \mathcal{T} \subset \mathcal{S} \wedge \sum_{s\in\mathcal{T}} \delta(s,t) \leq \tfrac{1}{2} \right\}$$

A low space conductance indicates that there are subsets with not much probability of containing the current state, and yet a small chance of leaving that subset soon after being trapped. A high space conductance indicates a high probability of getting out of such a set. The latter seems to be beneficial for fast convergence. We know from section 9.1 that there is a relation between convergence and the second eigenvalue of the transition matrix. We will now prove that the two values, $\Phi_{\mathcal{S},\mu}(t)$ and $\lambda_2(t)$ are related.

Lemma 10.1 **T** is the transition matrix of a reversible chain, **D** is a diagonal matrix with $d_{ii} = d_i = \delta(s_i, t)$. $\mathbf{a}_s$ is a vector such that

- $\mathbf{aDT} = \lambda_2 \mathbf{aD}$,
- states are numbered such that $\forall_{1\leq i<s}[a_i \geq a_{i+1}]$,
- if **a** has exactly r positive components then $\sum_{i=1}^{r} d_i \leq \frac{1}{2}$.

Another vector $\mathbf{b}_s$ is defined as

$$b_i = \begin{cases} a_i & \text{if } a_i > 0 \\ 0 & \text{if } a_i \leq 0 \end{cases} .$$

Then

$$\sum_{i=1}^{r} \sum_{j=i+1}^{s} d_i t_{ij} (b_i^2 - b_j^2) \leq \sqrt{2 - 2\lambda_2} \sum_{i=1}^{r} d_i a_i^2 ,$$

■

Proof: It is easily checked that there is an a that satisfies all the conditions of the theorem. So, we concentrate on the formula, and consider the two sums $\sum_{i=1}^{r}\sum_{j=i+1}^{s} f_{ij}(b_i - b_j)^2$ and $\sum_{i=1}^{r}\sum_{j=i+1}^{s} f_{ij}(b_i + b_j)^2$, where $f_{ij} = d_i t_{ij}$ ($f_{ij} = f_{ji}$, of course, because the chain is reversible).

$$\sum_{i=1}^{r}\sum_{j=i+1}^{s} f_{ij}(b_i - b_j)^2 =$$

$$\sum_{i=1}^{r}\sum_{j=i+1}^{s} f_{ij}(b_i^2 + b_j^2) - 2\sum_{i=1}^{r}\sum_{j=i+1}^{s} f_{ij}b_ib_j =$$

$$= \sum_{i=1}^{r}\sum_{\substack{j=1\\ j\neq i}}^{s} f_{ij}b_i^2 - \sum_{i=1}^{r}\sum_{\substack{j=1\\ j\neq i}}^{r} f_{ij}b_ib_j =$$

$$= \sum_{i=1}^{r} d_ib_i^2 \sum_{\substack{j=1\\ j\neq i}}^{s} t_{ij} - \sum_{i=1}^{r}\sum_{\substack{j=1\\ j\neq i}}^{r} b_id_it_{ij}b_j =$$

$$= \mathbf{bD(I-T)b}^{\mathsf{T}} \leq \mathbf{aD(I-T)b}^{\mathsf{T}} = (1-\lambda_2)\sum_{i=1}^{r} d_ia_i^2.$$

The inequality follows from the fact that all terms on the right hand side that are not included in the left hand side, are positive. Further, remember that $t_{ii} = 1 - \sum_{j\neq i} t_{ij}$. Also note that **aD** is the eigenvector associated with λ_2.

$$\sum_{i=1}^{r}\sum_{j=i+1}^{s} f_{ij}(b_i + b_j)^2 \leq 2\sum_{i=1}^{r}\sum_{j=i+1}^{s} f_{ij}(b_i^2 + b_j^2) =$$

$$= \sum_{i=1}^{r} d_ib_i^2 \sum_{\substack{j=1\\ j\neq i}}^{s} t_{ij} \leq 2\sum_{i=1}^{r} d_ia_i^2.$$

Combining the two results

$$1-\lambda_2 \geq \frac{\sum_{i=1}^{r}\sum_{j=i+1}^{s} f_{ij}(b_i - b_j)^2}{\sum_{i=1}^{r} d_ia_i^2} \cdot \frac{\sum_{i=1}^{r}\sum_{j=i+1}^{s} f_{ij}(b_i + b_j)^2}{2\sum_{i=1}^{r} d_ia_i^2} \geq$$

$$\frac{1}{2}\left(\frac{\sum_{i=1}^{r}\sum_{j=i+1}^{s} f_{ij}(b_i^2 - b_j^2)}{\sum_{i=1}^{r} d_ia_i^2}\right)^2$$

qed

Lemma 10.2 With the notations of lemma 10.1 an upper bound on the space conductance Φ can be written as

$$\frac{\sum_{i=1}^{r}\sum_{j=i+1}^{s} d_i t_{ij}(b_i^2 - b_j^2)}{\sum_{i=1}^{r} d_i a_i^2}.$$

■

Proof:

$$\begin{aligned}
\sum_{i=1}^{r}\sum_{j=i+1}^{s} f_{ij}(b_i^2 - b_j^2) &= \sum_{i=1}^{r}\sum_{j=i+1}^{s} f_{ij}\sum_{k=i}^{j-1}(b_k^2 - b_{k+1}^2) = \\
&= \sum_{k=1}^{r}(b_k^2 - b_{k+1}^2)\sum_{i=1}^{k}\sum_{j=k+1}^{s} f_{ij} = \\
&= \sum_{k=1}^{r}(b_k^2 - b_{k+1}^2)\sum_{i=1}^{k} d_i\frac{\sum_{i=1}^{k}\sum_{j=k+1}^{s} f_{ij}}{\sum_{i=1}^{k} d_i} \geq \\
&\geq \Phi\sum_{k=1}^{r}(b_k^2 - b_{k+1}^2)\sum_{i=1}^{k} d_i = \\
&= \Phi\sum_{i=1}^{r} d_i\sum_{k=i}^{r}(b_k^2 - b_{k+1}^2) = \\
&= \Phi\sum_{i=1}^{r} d_i(b_i^2 - b_{r+1}^2) = \Phi\sum_{i=1}^{r} d_i a_i^2
\end{aligned}$$

qed

The two lemmas yield straightforwardly the following theorem.

Theorem 10.1 The second eigenvalue of the transition matrix of a reversible chain and the space conductance satisfy

$$2\lambda_2 + \Phi^2 \leq 2.$$

■

Our intuition that a large space conductance is beneficial for the chain convergence is confirmed by this theorem, for the larger the space conductance the smaller the bound on the second eigenvalue.

10.2 The topography of the state space

To facilitate the description of the state space we will develop a suitable terminology in this section. Much of this terminology is inspired by viewing the state space as a landscape where distance is often roughly measured by the number of moves necessary to go from one state to another, and altitude is equivalent to score. The features described by this terminology are part of what is therefore called the *topography* of the state space. Special interest is in what can be said about the local minima in that context. Let us therefore first define the set of local minima. Since a state is called a *local minimum* if there is no state in $\mathcal{S}$ that is connected to that state via a single move and has a lower score, the set of local minima is properly defined as

$$\mathcal{L} = \{s \mid \forall_{s\prime \in \mathcal{S}\mu}[\varepsilon(s\prime) \geq \varepsilon(s)]\}$$

One way of defining a distance between two local minima is to take the minimum number of moves necessary to reach one from the other:

$$v(s, s') = \min\{\mathsf{k} \mid (s, s') \in \mu^{\mathsf{k}}\}.$$

Let us call this the *move distance.* This distance is completely independent of the score differences that have to be accepted when moving between these states. Intuitively one wants to relate distance also with score: large differences in score should require many moves. In other words, we want the landscape to be smooth. "Barriers" between local minima with only a few moves between them should not be much higher than these minima. We therefore want to introduce another distance between local minima, the *barrier distance,* that reflects the highest score that has to be encountered.

Before defining the barrier distance we first introduce the concept of clipping the set of states. The state set *clipped* at a certain level $\tilde{\varepsilon}$ is denoted by $\mathcal{S}_{\tilde{\varepsilon}}$:

$$\mathcal{S}_{\tilde{\varepsilon}} = \{s \mid \varepsilon(s) \leq \tilde{\varepsilon}\}$$

and the restriction of the move set to $\mathcal{S}_{\tilde{\varepsilon}}$ by $\mu_{\tilde{\varepsilon}}$:

$$\mu_{\tilde{\varepsilon}} = \mu \cap (\mathcal{S}_{\tilde{\varepsilon}} \times \mathcal{S}_{\tilde{\varepsilon}}).$$

$\mu_{\tilde{\varepsilon}}$ preserves symmetry and reflexivity, but may have los t the property that its transitive closure is the universal relation over $\mathcal{S}_{\tilde{\varepsilon}}$. The transitive closure is, of course, transitive, but also symmetric and reflexive, and therefore an equivalence relation. Or, in other words, it induces a partition over $\mathcal{S}_{\tilde{\varepsilon}}$. The blocks in that partition are the *dips* under $\tilde{\varepsilon}$. The barrier distance assigned to a pair of local minima is now defined as

$$\eta(s, s\prime) = \min\{\varepsilon \mid (s, s\prime) \in \bigcup_{\mathsf{k}=1}^{\infty} \mu_{\tilde{\varepsilon}}^{\mathsf{k}}\}$$

when $s \neq s\prime$ and $\eta(s, s) = 0$. The range of η is called the set of *rim levels* for that particular state set and score function.

The barrier function has the following properties:

$$\forall_{s\in\mathcal{S}}\forall_{s'\in\mathcal{S}}[0 \le \eta(s,s') < \infty]$$

$$\forall_{s\in\mathcal{S}}\forall_{s'\in\mathcal{S}}[\eta(s,s') = 0 \Leftrightarrow s = s']$$

$$\forall_{s\in\mathcal{S}}\forall_{s\prime\in\mathcal{S}}[\eta(s,s') = \eta(s',s)]$$

$$\forall_{s\in\mathcal{S}}\forall_{s'\in\mathcal{S}}\forall_{s^o\in\mathcal{S}}[\eta(s,s') \le \eta(s,s^o) + \eta(s^o,s')]$$

and therefore $(\mathcal{L},\eta)$ is a metric space. But η satisfies an even stronger property:

$$\forall_{s\in\mathcal{S}}\forall_{s'\in\mathcal{S}}\forall_{s^o\in\mathcal{S}}[\eta(s,s') \le \max(\eta(s,s^o),\eta(s^o,s'))]$$

which makes $(\mathcal{L},\eta)$ an *ultrametric space.* It is easy to show that the last property is equivalent to the requirement that for each triple of local minima two of their mutual barrier distances are equal to each other, and the third one is smaller than those two.

$$\forall_{s\in\mathcal{S}}\forall_{s'\in\mathcal{S}}\forall_{s^o\in\mathcal{S}}[\eta(s,s') < \eta(s,s^o) \Rightarrow \eta(s,s^o) = \eta(s^o,s')]$$

Every finite ultrametric space can be represented by a tree, and conversely every tree with a certain level numbering induces an ultrametric space. Also $(\mathcal{L},\eta)$ can therefore be represented by a tree. The vertices of this tree are subsets of $\mathcal{L}$. The leaves are the subsets with only one element of $\mathcal{L}$. The other vertices represent maximal subsets $\mathcal{L}_b$ with the property that

$$\forall_{(s,s')\in\mathcal{L}_b\times\mathcal{L}_b}[\eta(s,s') \le b].$$

For each subset $\mathcal{L}_b$ a suitable rim level can be chosen to replace b. This number is also assigned to the associated vertex. The tree can be seen as a hierarchy for the local minima.

With every pair in $(s,s') \in \mathcal{L}\times\mathcal{L}\backslash\iota$ we also associate a set of states that we will call its *valley*:

$$\mathcal{V}(s,s') = s\bigcup_{k=1}^{\infty}\mu_{\tilde{\varepsilon}}^{k}$$

where $\tilde{\varepsilon} = \min\{\eta(s,s^o) | s^o \in \mathcal{L} \wedge (\eta(s,s^o) > \eta(s,s\prime))\} - \frac{1}{2}d\varepsilon$, and $d\varepsilon = \min\{|\varepsilon(s_i) - \varepsilon(s_j)| \, |\varepsilon(s_i) \ne \varepsilon(s_j)\}$.

Further, $\mathcal{V}(s,s) = \{s\}$. A valley is therefore the smallest dip in which the two minima can reached from one another. If the intersection two valleys is not empty then one of these valleys is contained in the other. If we construct a digraph with the valleys of $\mathcal{S}$ as its vertices, and with an arc from $\mathcal{V}$ to $\mathcal{V}' \subset \mathcal{V}$ if there is no $\mathcal{V}^l$ such that $\mathcal{V}' \subset \mathcal{V}^l \subset \mathcal{V}$, then we obtain a tree that is isomorphic with the hierarchy for local minima.

So, the set of local minima and the barrier distance always form an ultrametric space. If high barriers between local minima should imply a large move distance, then $(\mathcal{S}, v)$ should also "tend to ultrametricity", especially between the local minima with relatively low scores. This has led to the conjecture that state spaces for which $(\mathcal{S}, v)$ is almost ultrametric are well-structured for annealing, while large deviations from ultrametricity indicate poor behavior of annealing algorithm. The degree of ultrametricity can be measured by the correlation between the two larger distances in each triple of local minima. Experiments show that state spaces with which consistent success has been achieved with annealing, do show a strong correlation.

A strong correlation between score difference and move distance also translates the intuition that deep valleys should also be big valleys, while small valleys should be shallow. The relation with the space conductance of the preceding section suggests itself in this context!

10.3 The set of moves

It should be clear by now that constructing a state space for annealing requires careful consideration. It is not difficult to come up with really bad move sets. For example, putting all states into an arbitrary sequence and only allowing moves that either leave the state unchanged or move to a state immediately before or after the present state in the sequence, satisfies all the requirements, but has no advantage over a random search. Also, connecting all states with each other by a move, gives a valid move set, but is equivalent to randomly and uniformly selecting states from $\mathcal{S}$ as candidates for the next state. Both examples have all the properties that were established in the preceding chapters, but are obviously not efficient ways to obtain good solutions.

Of course, not all properties of the state space are affected by the move set. The equilibrium function, for example, does not depend on the move set under the accepted constraints, as we saw in chapter 5. Also, the global minima are completely determined by the score function. The local minima, however, do depend on the move set, and together with the barriers in between them they have big influence on the convergence of the chains. The two examples given in the previous paragraph are in that sense two extremes. If each state is connected with every other state the only local minima will be the global minima. If the move set is based on a sequence as described in previous paragraph there may be up to $\frac{1}{2}|\mathcal{S}| + (|\mathcal{S}| \mod 2)$ local minima. The former has the smallest possible number of local minima while the other may have the maximum number of local minima. These two examples are also extremes in space diameter. One has the minimum diameter (1), and one has the maximum diameter ($|\mathcal{S}| - 1$). Intuitively, a small diameter and few local minima seem to be desirable properties, but they also imply large score changes over a single move, and that does not seem to be desirable. Clearly, a good move set is a compromise.

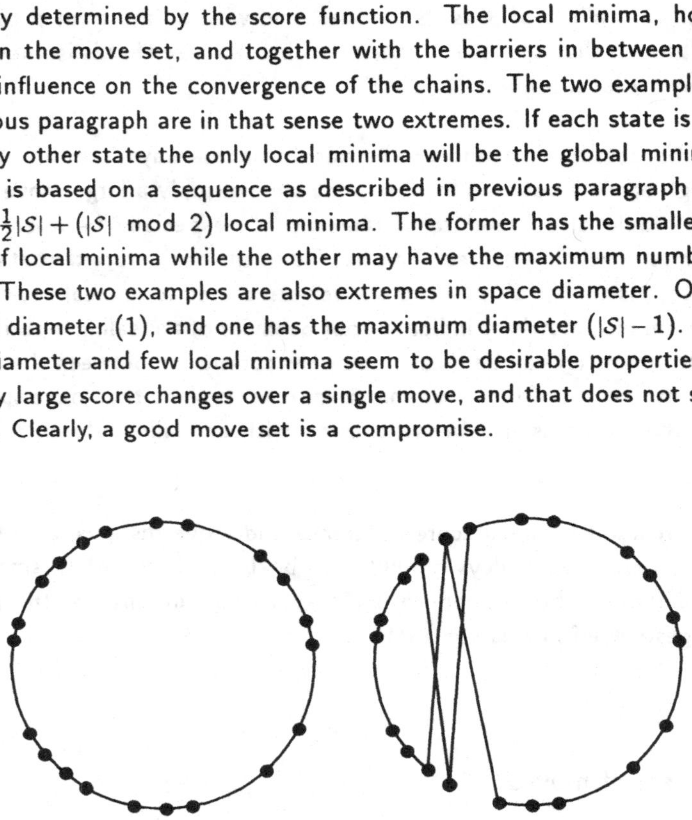

Figure 10.1: A move for the traveling salesman problem can be interchanging two cities.

The quality - or lack of quality - of the chosen move set is not always that obvious. An interesting example in that context is the traveling salesman problem. It was described in the first chapter of this book. Several move sets can be chosen. The first one may think of is called *swapping*: selection consists of producing

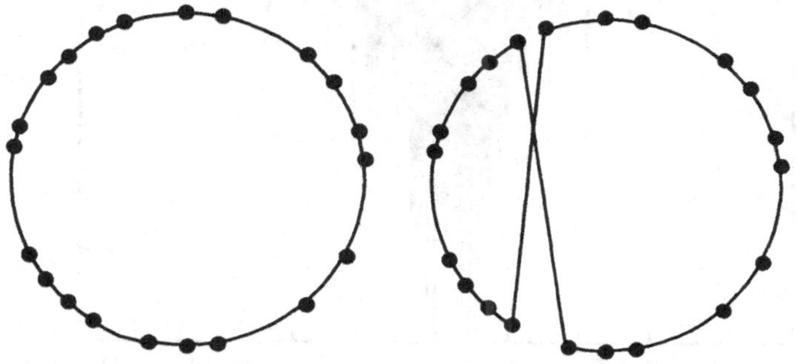

Figure 10.2: A better move for the traveling salesman problem is reversing part of the tour.

a random pair of distinct cities, and interchange their positions in the sequence (Figure 10.3). Another possibility is to select a subset of subsequent cities in the current tour and to reverse their sequence. This is called *reversal*. It is usually described as selecting two *bonds* (a bond is the connection between two subsequent cities in the tour) and replacing them by two new bonds such that a new tour is obtained (Figure 10.2). Results obtained with reversals as moves are generally much better than those with swapping. To get an idea why, a closer look at the respective landscapes is worthwhile.

To give a visual impression of the landscapes *space plots*, a kind of bar charts, were made with the states suitably ordered along the horizontal axis, and score levels along the vertical axis. By darkening the vertical bar of a state above its score value a clear picture of the valley structure is obtained. The states in a dip under $\tilde{\varepsilon}$ form a continuous dark line segment on the score level $\tilde{\varepsilon}$. That such state orderings exist follows from the tree structure described in section 10.2. To construct such an ordering we start by identifying all local minima with respect to the move set. A branch and bound technique was used to do an exhaustive search of the state space. The search was reduced by using the fact that a section of a locally optimal tour cannot be improved by performing moves within that section. That means that a partial solution cannot be improved by swaps or reversals that change the bonds within that section.

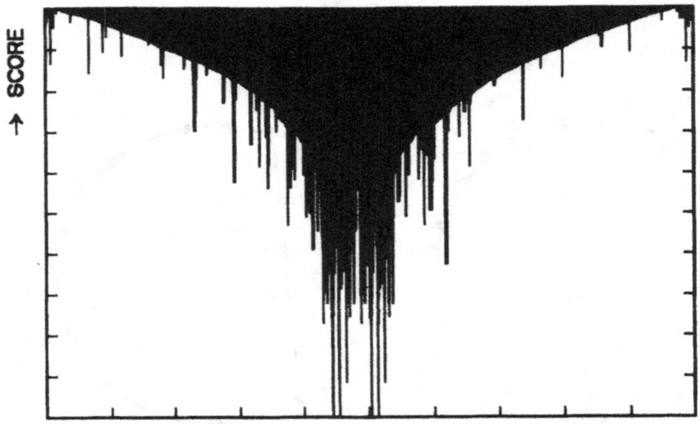

Figure 10.3: Space plot for the traveling salesman problem with swapping.

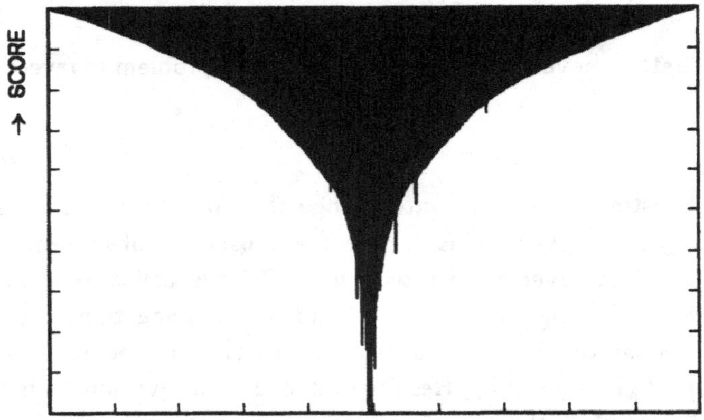

Figure 10.4: Space plot for the traveling salesman problem with reversals.

The space plots for the two move sets show a clear difference (Figure 10.3 and 10.4). The space in which the moves are swaps, has lots of local minima, quite deep and narrow. The space with reversals, on the other hand, has a much smoother appearance: only a small number of local minima and not as deep as many of the local minima in the other space. After seeing such a picture nobody will be surprised that reversals as moves yield a better space for annealing than the one constructed by swapping.

We also consider the minimum dependence period for these two spaces. In section 9.2 we saw that P can be used as an indicator for the convergence speed. The rate of convergence decreases as the size of the problem increases. It is therefore to expected P to grow with the number of cities. If there are N cities in the tour, the number of possible directed tours is the number of states $|\mathcal{S}| = (N-1)!$. Initially all tours are equally likely, so, using Stirling's formula, $H_\infty \approx (N - \frac{1}{2})\ln(N-1) - N + 1 + \ln(2\pi)/2$. The two cities to be swapped or the two bonds to be broken are selected from a uniform distribution making $\beta = 2(N-1)^{-2}$. Therefore initially $h_\infty = 2\ln(N-1) - \ln(2)$. The dependence period grows approximately with the number of cities as

$$P_\infty = \frac{(N-\frac{1}{2})\ln(N-1) - N + 1 + \ln(2\pi)/2}{2\ln(N-1) - \ln(2)} \approx \frac{1}{2}N\left(1 - \frac{1}{\ln(N)}\right) \qquad (10.1)$$

which is almost linear.

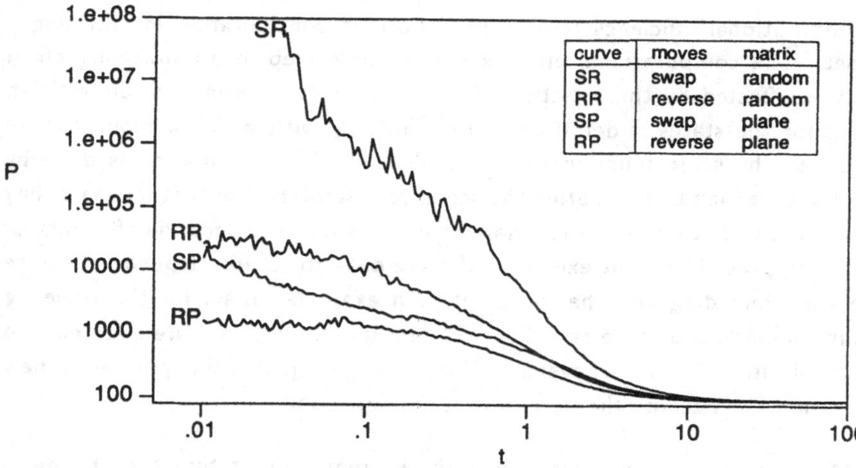

Figure 10.5: Behavior of different TSP problems

In figure 10.5 P is measured as a function of t for the four combinations of two move sets and two instances. The two instances have both 100 cities, but the first instance has symmetric, uniformly random distances, whereas the second instance has cities that are uniformly distributed in a plane square.

In this figure 10.5 we see that the minimum dependence period increases as t decreases. As expected, problems with euclidean distances in the plane are easier than problems with random distances. Also a move set of reversals gives a better convergence than swaps. Even though P is somewhat 'short sighted' as an indicator, the differences are very significant.

A space can also be constructed with a mixture of different types of moves. In a placement problem one may employ translations, pairwise interchanges, reflections, and others in the same space. A new type of move may serve as a catalyser. An interchange for example can also be achieved by two translations, but when overlap is heavily "punished" in the score function, the intermediate state forms a "high barrier". Such a high intermediate, with a low acceptance probability for the lower values of t, can be avoided by an interchange, and offer direct moves with a much smaller score difference, and thus a higher acceptance probability. These so-called catalytic moves can smooth the landscape quite a bit, and improve convergence in this way.

The computational efficiency is also an important consideration in choosing a move set. The computational efficiency of a single step in an annealing chain certainly is affected by the selection of the move set. In combination with an encoding for the states it determines the efficiency with which a move can be effected and the score function can be evaluated. This evaluation is done by updating rather than by computing the score from scratch. For that reason it may be convenient to have intermediate data structures that enhance the efficiency of the updating procedure. The existence of these data structures depends of course on the state encoding and the move set. An example, again for the traveling salesman problem, is a move set where not two but three bonds are selected. No sequence of cities has to be reversed. With changing just a few pointers a new configuration has become the new current configuration.

Efficient computation often coincides with the requirement that the change in object function be not too large. In the case that the object function can be described as the sum of more or less independent costs then it is often efficient to have a move set that changes as few as possible terms of this sum.

10.4 Global convergence

One important question remained unanswered in this book: Is there a schedule, that is a sequence of values for the control parameter, such that with probability 1 the final state is a global minimum. We mentioned in section 9 that it would require infinitely many steps, but we did not give a proof of that statement. Also here we are not going to prove the corresponding theorem, but we do want to phrase that theorem in our terminology, because the implication that a global minimum beyond any doubt is not reachable in finite time is of course valuable.

Theorem 10.2 (Hajek) The annealing algorithm with t_k as the value of the control parameter in the kth step produces a global minimum state with probability 1 fif

$$\sum_{k=1}^{\infty} \exp\left(-\frac{h}{t_k}\right) = \infty$$

where h is the smallest number such that

$$\forall_{s \in \mathcal{L} \setminus S_0} \exists_{s' \in \mathcal{L}} [\varepsilon(s') < \varepsilon(s) \wedge \eta(s, s') = \varepsilon(s) + h]$$

■

An example of such a schedule is one with

$$t_k = \frac{c}{\ln(k+1)} \tag{10.2}$$

with $c \geq h$. Like any schedule guaranteeing a globally optimum state in an annealing space with at least one non-global local minimum it requires an infinite number of steps.

10.5 Discussion

The concept of space conductance as well as its relation with the second eigenvalue of the transition matrix are due to [139]. Ultrametricity is a concept that due to recent developments in the mean field theory of spin glasses[149], became known to people interested in annealing for combinatorial optimization [84], [143]. Catalytic moves were so named in [70], where also their effect was described.

The major result on the global convergence of the annealing algorithm is due to Bruce Hajek who announced and distributed the result already in 1985. However, the official publication did not appear till May 1988 [60]. Other results were obtained around the same time. [47] was the first with a schedule as given in (10.2), though the numerator was chosen quite generously. [108] came with sharper bounds and a similar schedule, as did [4]. [46] had a more general result in that they showed that convergence to an arbitrary subset can be ensured by such a schedule. Also [60] had a more general result than theorem 10.2 suggests. The annealing chains in this book are assumed to be reversible. [60] only requires weak reversability: s is reachable from s' via states with scores less than $\tilde{\varepsilon}$ fif s' is reachable from s via states with scores less than $\tilde{\varepsilon}$.

11 IMPLEMENTATION ASPECTS

So far this book described the annealing algorithm in general terms of states, moves and scores. For an implementation, these concepts have to be associated with concrete configurations, modifications and objective functions in such a way that they can be manipulated by a computer, i.e. written in a programming language. For many problems this task is relatively straightforward. The state space and the score function are usually strongly suggested by the problem specification. The move set should be chosen such that the modifications and new scores can be computed fast, and a smooth space landscape is realized.

In this chapter we want to give the annealing implementation of a specific optimization problem: placement of modules in an integrated circuit. After a short statement of the problem a listing is included of an ANSI-Pascal program for annealing applied to that problem. In the other two sections of this chapter some ways for speed-up are discussed. First some more general methods based on modifying the move set, followed by speed-ups that are more application dependent.

11.1 An implementation

One of the most successful applications of annealing, and also the one that illustrated the introduction of the concept in the field of optimization [82],[83], is the placement of modules in an integrated circuit. These modules are seen as squares of equal size that can occupy slots in a rectangular array of squares. They form a set in an incidence structure as introduced in chapter 3. Modules that are "incident" with the same net have to be connected by signal wires. These wires are to be kept as short as possible for several reasons, the two most important ones being area taken up by these wires and signal delay.

To apply annealing the following state space and score function have been chosen. The states correspond with the different possible placements of the modules. The positions of the modules characterize the state. The score function will have to measure the amount of wiring. We can make an estimate of the length of the wires by using the bounding box method: the length of a wire is half the perimeter of the smallest box enclosing all modules incident with the same net representing that wire. The score of a configuration is the sum of the lengths of the wires. To move from one state to another the positions of two modules are interchanged.

In the program the current state is kept in the variable `cstate`. This variable gives the position of each module. For efficiency reasons, this state is also kept in the variable `chip` which gives the module of each position. These variables can be seen as each others inverse. `chip` is built from `cstate` by calling `restart`.

The net list is stored in the array `mod2net`. This array contains the net numbers, sorted by module number. `mods[m].netpt1` and `mods[m].netpt2` are pointers in this array, indicating the first and last net connected to module m. For computing the score update efficiently, the 'inverse' of the net list is stored in the array `net2mod`. This array contains the module numbers, sorted by net number. `nets[n].modpt1` and `nets[n].modpt2` point to the first and last module connected to net n. `net2mod` is built from `mod2net` by calling `buildnet2mod`.

The random moves are selected and performed by the procedure `move`. If rejected, the move can be undone by calling `undomove`. The score of the current state is kept in the variable `netlength`. This variable is initialized by calling `restart`. Rather than recomputing this variable from scratch after every move, it is updated. This is done in the procedure `transfer`. Only the lengths of the nets involved in the move are considered. The range of each net is saved in the array `nets[n].range`, and can be computed by procedure `netrange` .

The schedule control is implemented according to the ideas that have been developed in this book. The procedure `initialization` determines two of the three parameters of the schedule E_∞ and σ_∞. The third one is determined in `metropolis`, which executes the inner loop.

```
PROGRAM gatearray(input, output, netlist, placement);

{     Gate array placement by simulated annealing.

      The first line in the input file gives the X and Y dimensions of
      the gate array. Each further line represents the pins of one
      module:  each line contains the net numbers of the nets
      connecting to that module. The net numbers must be consecutive,
      i.e. all net numbers must occur at least twice.

      The output file has two columns. The first is the X-coordinate,
      the second the Y-coordinate of the modules, in the same sequence
      as in the input file.
}


CONST mmax   =  100;                   { maximum number of modules }
      nmax   =  500;                   { maximum number of nets }
      lmax   = 1000;                   { maximum number of pins }
      xmax   =   20;                   { maximum dimension of chip }


TYPE  axis      =   (x, y);            { x axis and y axis }
      position  =   ARRAY [axis]       { pair of coordinates }
                          OF integer;
      space     =   ARRAY [1..Mmax]    { the coordinates of the module }
                          OF position;


VAR   mods: ARRAY [1..mmax] OF         { list of modules, mdim is last }
            RECORD netpt1,             { pointers to mod2net }
                   netpt2: integer;    { to find connected nets }
            END;
      mod2net: ARRAY [1..lmax]         { net numbers ordered by module }
                  OF integer;
      nets: ARRAY [1..nmax] OF         { the list of nets, ndim is last }
      RECORD modpt1,                   { pointers to net2mod }
             modpt2: integer;
             rrange: ARRAY [axis] OF   { net range for each direction }
                   RECORD from,        { lowest coordinate }
                          too: integer; { highest coordinate }
                   END;
      END;


      net2mod: ARRAY [1..lmax]         { module numbers ordered by net }
                  OF integer;          {   (inverse of mod2net) }
      cscore,                          { score of the current state }
      bestscore: real;                 { best score encountered so far }
      beststate,                       { best state encountered }
      cstate: space;                   { current state }
      chip: ARRAY [1..xmax, 1..xmax]   { chip image (inverse of cstate) }
               OF integer;
      mdim: integer;                   { number of modules }
      ndim: integer;                   { number of nets }
      ldim: integer;                   { number of pins }
      xdim, ydim: integer;             { dimensions of chip image }
      netlength: real;                 { sum of net lengths }
      netlist, placement: text;        { input and output file }
      p1, p2: position;                { positions involved in a move }
```

```
FUNCTION random: real; EXTERNAL;       { random number in interval [0,1> }

PROCEDURE load;                        { load net list file }
VAR   i, m: integer;
BEGIN reset(netlist);
      readln (netlist, xdim, ydim);
      mdim := 0;
      ldim := 0;
      WHILE NOT eof (netlist) DO
      BEGIN mdim := mdim + 1;
            mods[mdim].netpt1 := ldim + 1;
            WHILE NOT eoln (netlist) DO
            BEGIN ldim := ldim + 1;
                  read (netlist, mod2net [ldim]);
            END;
            readln (netlist);
            mods[mdim].netpt2 := ldim;
      END;
      writeln('Number of modules = ', mdim:1);
END; { load }


PROCEDURE save;                        { save best placement on disk }
VAR   i: integer;
BEGIN rewrite(placement);
      FOR i := 1 TO mdim
         DO writeln (placement, beststate[i,x]:6, beststate[i,y]:6);
END; { save }


PROCEDURE buildnet2mod;                { build inverted net list }
VAR   i, k, m: integer;
      found: boolean;
BEGIN ndim := 0; k := 0;
      REPEAT found := false;
             ndim := ndim+1;
             nets[ndim].modpt1 := k+1;
             FOR m := 1 TO mdim DO
                FOR i := mods[m].netpt1 TO mods[m].netpt2 DO
                   IF mod2net[i] = ndim THEN
             BEGIN found := true;
                   k := k + 1;
                   net2mod[k] := m;
                   nets[ndim].modpt2 := k;
             END;
             IF nets[ndim].modpt2 = nets[ndim].modpt1
             THEN writeln('Warning: dangling net ', ndim:1);
      UNTIL NOT found;
      ndim := ndim-1;
      writeln('Number of nets    = ', ndim:1);
END; { buildnet2mod }


PROCEDURE netrange(n: integer; ax: axis); { compute the range of a net }
VAR   i, pos: integer;
BEGIN WITH nets[n], rrange[ax] DO
      BEGIN from := xmax;
            too := 0;
            FOR i := modpt1 TO modpt2 DO
            BEGIN pos := cstate[net2mod[i],ax];
                  IF pos > too  THEN too := pos;
                  IF pos < from THEN from := pos;
            END;
      END;
END; { netrange }
```

```
PROCEDURE transfer(m: integer; p: position);
      { transfer module m to position p }
VAR   i: integer;
      ax: axis;
BEGIN cstate[m] := p;
      FOR i := mods[m].netpt1 TO mods[m].netpt2 DO
         FOR ax := x TO y DO
            WITH nets [mod2net[i]], rrange[ax] DO
      BEGIN netlength := netlength - (too-from);
            netrange (mod2net[i], ax);
            netlength := netlength + (too-from);
      END;
END; { transfer }


PROCEDURE swap (p1, p2: position);
      { swap the contents of positions p1, p2 }
VAR   m1, m2: integer;
BEGIN m1 := chip [p1[x], p1[y]];
      m2 := chip [p2[x], p2[y]];
      IF m1 <> 0 THEN transfer(m1, p2);
      IF m2 <> 0 THEN transfer(m2, p1);
      chip [p2[x], p2[y]] := m1;
      chip [p1[x], p1[y]] := m2;
END; { swap }


PROCEDURE move;        { select and perform a random move }
BEGIN p1 [x] := trunc (random * xdim) + 1;
      p1 [y] := trunc (random * ydim) + 1;
      p2 [x] := trunc (random * xdim) + 1;
      p2 [y] := trunc (random * ydim) + 1;
      swap(p1, p2);
END; { move }


PROCEDURE undomove;                        { undo the previous move }
BEGIN swap(p1, p2) END;

FUNCTION identmove: boolean;               { test for identity moves }
BEGIN identmove := (p1[x] = p2[x]) AND (p1[y] = p2[y]); END;


PROCEDURE restart;          { recompute the net length from scratch }
VAR   i, j: integer;
      ax: axis;
BEGIN FOR i := 1 TO xdim DO
         FOR j := 1 TO ydim
            DO chip[i,j] := 0;
      FOR i := 1 TO mdim
         DO chip[cstate[i,x], cstate[i,y]] := i;
      netlength := 0;
      FOR i := 1 TO ndim DO
         FOR ax := x TO y DO
      BEGIN netrange(i, ax);
            WITH nets[i].rrange[ax] DO
            netlength := netlength + (too - from);
      END;
END; { restart }
```

```
PROCEDURE generate;                        { generate a random state }
VAR   i, h: integer;
      tmp: position;
BEGIN FOR i := 1 TO xdim * ydim DO
      BEGIN cstate[i, x] :=( i-1) MOD xdim + 1;
            cstate[i, y] := (i-1) DIV xdim + 1;
      END;
      FOR i := xdim * ydim DOWNTO 1 DO
      BEGIN h := trunc(random * i) + 1;
            tmp := cstate[h];
            cstate[h] := cstate[i];
            cstate[i] := tmp;
      END;
      restart;
END; { generate }

FUNCTION epsilon: real;                    { the score or objective function }
BEGIN epsilon := netlength END;

FUNCTION beta: real;                       { move selection probability }
BEGIN beta := 2.0 / sqr(xdim*ydim); END;

FUNCTION stirling(x: real): real;          { Stirling's formula }
CONST pi = 3.1415926536;
BEGIN stirling := (x+0.5)*ln(x) - x + ln(2*pi)/2; END;

FUNCTION Hinf: real;                       { initial accessibility }
BEGIN Hinf := stirling(xdim*ydim) - stirling(xdim*ydim - mdim);
END; { Hinf }

PROCEDURE selfcheck;                 { check for bugs in update mechanism }
VAR   oldscore: real;
BEGIN oldscore := epsilon;
      restart;
      IF epsilon <> oldscore
         THEN writeln('Bug in score update.');
END; { selfcheck }

PROCEDURE anneal;                    { adaptive annealing schedule }
CONST gamma    =   0.1;              { change in distribution per t step }
      c        =     4;              { multiplier for inner loop }
      r        =   384;              { number of independent samples }
      theta    =  0.01;              { 1% precision of optimum }
      maxmoves = 10000;              { max number of moves per t step }
VAR   t,       {t=9999..}            { current control parameter }
      H,             H99,            { global accessibility }
      E,             E99,            { average score }
      hl,            hl99,           { local accessibility }
      sigma,   sigma99,              { standard deviation }
      dt,                            { change in control parameter }
      TT,                            { border between strong and weak control}
      te,                            { lower end of strong control region }
      mu,                            { smoothing estimate }
      maxte: real;                   { maximum of estimates of te }
      k: integer;                    { inner loop count }

 FUNCTION expon(x: real): real;            { protected exp(x) function }
 BEGIN IF x < -40
       THEN expon := 0
       ELSE IF x > 40
            THEN expon := exp(40)
            ELSE expon := exp(x);
 END; { expon }
```

```
PROCEDURE initialization;              { find E99, H99, sigma99, hl99, t }
VAR   i: integer;
      sum, sum2: real;                 { collected statistical data }
BEGIN sum  := 0;
      sum2 := 0;
      FOR i := 1 TO r DO
      BEGIN generate;
            sum  := sum  + epsilon;
            sum2 := sum2 + sqr(epsilon);
      END;
      E99 := sum / r;
      sigma99 := sqrt( (sum2 - sqr(sum)/r ) / (r-1) );
      t := sigma99 * sqrt(1 / gamma);
      H99 := Hinf;
      hl99 := -ln(beta);
      sigma := sigma99;
      E := E99;
      H := H99;
      hl := hl99;
END; { initialization }


PROCEDURE metropolis(t: real);         { metropolis loop }
VAR   sum: real;                       { sum of scores }
      Eold: real;                      { previous average score }
      history: ARRAY[1..maxmoves]      { save all scores encountered }
         OF real;
      variance,                        { used in computing sigma }
      nscore: real;                    { proposed new score }
      a,                               { probability of state changes }
      up: real;                        { average upward score change }
      xlnx: real;                      { x.ln(x) = 0 if x=0 }
      i: integer;


BEGIN Eold := E;
      a := 0;
      up := 0;
      FOR i := 1 TO k DO               { inner or metropolis loop }
      BEGIN move;
            nscore := epsilon;
            IF random < expon (-(nscore - cscore) / t) THEN
            BEGIN IF NOT identmove THEN a := a + 1/k;
                  IF cscore < nscore
                     THEN up := up + (nscore-cscore)/k;
                  cscore := nscore;
            END
               ELSE undomove;
            IF cscore < bestscore THEN
            BEGIN bestscore := cscore;
                  beststate := cstate;
            END;
            history[i] := cscore;
      END;
      sum := 0;
      FOR i := 1 TO k
         DO sum := sum + history[i];
      E := sum / k;
      H := H - (Eold - E) / t;
      sum := 0;
      FOR i := 1 TO k             { history avoids loss in precision }
         DO sum := sum + sqr(history[i] - E);
      variance := sum / (k-1);
      sigma := sqrt(variance);
```

```
      IF a = 1
         THEN xlnx := 0
         ELSE xlnx := (1-a)*ln(1-a);
      hl := (hl + a*ln(1/beta) + up/t - xlnx) / 2; { some smoothing }
      IF H * c > hl * maxmoves          { protect against too small hl }
         THEN hl := H * c / maxmoves;
      IF k < H/hl * c             { don't use fewer moves then before }
         THEN k := round(H/hl * c);
      selfcheck;
END; { metropolis }

FUNCTION smooth(z: real): real;         { exponential smoothing }
CONST omega = 0.90;                     { decay constant of smoothing }
BEGIN mu := (1-omega) * z + omega*mu*t / (t+dt);
      smooth := mu;
END; { smooth }

BEGIN initialization;
      cscore := epsilon;
      k := round(H99/hl99 * c);
      beststate := cstate;
      bestscore := cscore;
      mu := sigma99;
      dt := 0;
      maxte := 0;
      REPEAT writeln('t= ', t:6:2, ',  E= ', E:6:2, ',  H= ', H:7:2,
                 ',  sigma= ', sigma:5:2, ',  h= ', hl:6:2, ',  k= ', k:3);
           metropolis(t);
           sigma := smooth(sigma);
           IF 2 * gamma * t > sigma
              THEN dt := t / 2
              ELSE dt := gamma * sqr(t) / sigma;
           TT := sigma99 / sigma * t;
           t := t - dt;
           te := TT * expon(-sqr(TT) * H99 / sqr (sigma99) + 0.5);
           IF te > maxte THEN maxte := te;
      UNTIL (sqr(sigma) < theta * t * (E99-E)) { sigma small enough }
        AND (t < te);                          { end of linear area }
      metropolis(t/100);                { final iterative improvement }
      writeln('Best score = ', bestscore:6:2);
END; { anneal }

BEGIN writeln('    *** Gate Array Placement ***');
      load;
      buildnet2mod;
      anneal;
      save;
      writeln('    *** Placement Completed ***');
END. { gatearray }
```

11.2 The selection function

A small space diameter is desirable for having all states easily reachable. Yet the difference between the scores of two states connected by a move that can be selected, has to be relatively small. The choice of a move set and the selection function should therefore be a compromise between a small space diameter and a smooth score variation. Which compromise is best also depends on the value of the control parameter. A small space diameter is especially desirable in the first stages of annealing, while small score differences are necessary when t is low.

It must be clear, certainly after reading chapter 9 and 10, that the move set and the selection function do have a considerable effect on the efficiency of the annealing algorithm. The equilibrium density on the other hand was shown to be independent of the selection function under the accepted constraints (chapter 5). This immediately introduces the idea of adapting that function to the situation to improve the convergence properties of the chain.

The first idea is of course, to try to avoid selecting moves that almost certainly are going to be rejected. Of course, this only makes sense if such moves can be identified more efficiently than the acceptance probability can be evaluated. This is, for example, the case when for each move $(s,s') \in \mu$ a move length can be determined relatively quickly. $\ell(s,s')$ is called a move length when it is (almost) monotonically related to the average change in score over all moves with that length:

$$\overline{\Delta\varepsilon}(\mathrm{x}) = \frac{\Sigma\{|\varepsilon(s) - \varepsilon(s')| \,|\, (s,s') \in \mu \wedge \ell(s,s') = \mathrm{x}\}}{|\{(s,s') | (s,s') \in \mu \wedge \ell(s,s') = \mathrm{x}\}|}$$

Moves with a low acceptance probability can be avoided by limiting the move length of the selected moves, for example $\ell < \mathrm{L}$ with

$$\mathrm{L} = \max\left(\{\ell_{\min}\} \cup \{\ell | \overline{\Delta\varepsilon}(\ell) \leq -\mathrm{t}\ln(\zeta)\}\right).$$

A table representing the function $\overline{\Delta\varepsilon}(\ell)$ can be assembled during the initialization, and updated while running the algorithm by extracting the move length of each selected move and evaluating the score difference.

A more elaborate adaptive method to improve chain convergence by modifying the selection function was proposed in [70] where a move size is defined and used to assign a *quality* to each move. This number should express the fact that one rather has a move with a large score change and yet a reasonable acceptance probability.

One therefore assigns a *size* to each move (this might be the score change, its square, etc.). For each type of move (a type is the kind of transformation the current configuration has to undergo to become the next configuration) the product of the average size of an accepted move of that type and the acceptance ratio is its quality factor. This is done over a whole chain until the control parameter is changed. At this next value the selection probability of the moves is taken proportional to their quality. To make the technique more effective it is combined with *windowing*, thus generating a whole hierarchy of moves. An undesirable bias towards certain moves must be avoided.

In [54] examples were given in which from a certain *crossover point* on rejection of moves is avoided altogether. The method works as follows. For each move out of the current state, $(s, s') \in \mu$ $w_S(s') = \alpha(\varepsilon(s), \varepsilon(s)', t)$ is stored. The stored value is the probability that a move to s' is selected. This selected move is unconditionally accepted, and all $w_{s'}$ are stored. The problem to select a move with that probability and to update the w-values is called the *dynamic weighted selection problem.*

This modification does change the equilibrium density, because not only the selection function, but also the acceptance probability is changed. The succession of distinct states is "probabilistically" unchanged, and giving each state a weight corresponding with the expected number of repetitions yields the familiar equilibrium density again. Once the relative number of acceptances drops below a certain value, the crossover point, the rejectionless method becomes faster, provided we can solve the dynamic weighted selection problem efficiently for the problem at hand.

11.3 Other speed-up methods

In an attempt to reduce computation times required by some implementations of annealing a two-stage approach has been suggested. In the first stage a reasonably fast heuristic finds a "better-than-random" configuration. The second stage is annealing with the result of the first stage as initial configuration. The question that has to be answered then is what initial value the control parameter should get. The initialization of section 8.1 is certainly inadequate, because that is chosen with the intention of making the annealing as fast as possible independent of the initial state. The effect of the heuristic would be null in that case. Too low a value would make it difficult to escape from a nearby local minimum. If the implementation is used for instances of one and the same optimization problem the control parameter can be set according to previous experience [59]. In an implementation with a more general purpose intention this is not feasible. Not dependent on the specific application would be the answer to the question: for which value of t is the expected score E(t) equal to the score of the configuration produced by the fast heuristic. But that is exactly what annealing down to that value of the control parameter is supposed to do: bringing the chain in equilibrium at that, a priori unknown, value of the control parameter. Nevertheless, successful experiments with such an approach have been reported[127]. The procedure followed there is as follows. First the probability density function of $\Delta\varepsilon$ is measured. This is done however by randomly selecting states in $s_h\mu$ where s_h is the state corresponding with the result of the fast heuristic. Only score differences of moves starting with s_h are considered. A first value for the control parameter is guessed. Then depending on whether

$$\sum_{\Delta\varepsilon>0} \Delta\varepsilon \mathrm{pdf}(\Delta\varepsilon) \exp(-\frac{\Delta\varepsilon}{t}) > \sum_{\Delta\varepsilon<0} \Delta\varepsilon \mathrm{pdf}(\Delta\varepsilon)$$

is true or false, a higher or lower t is tried next. This iteration is stopped when

$$\sum_{\Delta\varepsilon>0} \Delta\varepsilon \mathrm{pdf}(\Delta\varepsilon) \exp(-\frac{\Delta\varepsilon}{t}) \approx \sum_{\Delta\varepsilon<0} \Delta\varepsilon \mathrm{pdf}(\Delta\varepsilon).$$

It is very questionable whether the measured density of $\Delta\varepsilon$ is representative for the real density function. Support is found in the fact that good results were obtained with one application: standard cell placement with min-cut as the first stage. The reason of the success may well be hidden in the special character of the heuristic: the heuristic finds a rough structure while the finer structure is not considered by the first stage. Annealing finds a finer structure and is kept from making big changes because of the relatively low t.

Another proposal for speeding up annealing is based on the observation that an error in the score difference that is small compared with the value of the control parameter does not have much influence. This brought the idea of *approximate calculations* [59] in which the effect of a move is mostly not determined exactly. An accurate update is only necessary when the total error is of the same order of of magnitude as t.

REFERENCES

[1] D.H. Ackley, G.E. Hinton, and T.J. Sejnowski. A learning algorithm for boltzmann machines. *Cognitive science: a multidisciplinary journal of artificial intelligence*, 9:147–169, 1985.

[2] S. Anily and A. Federgruen. *Probabilistic analysis of simulated annealing methods.* Technical Report, Graduate School of Business, Columbia University, New York, 1985. preprint.

[3] S. Anily and A. Federgruen. Simulated annealing methods with general acceptance probabilities. 1985. preprint, Graduate School of Business, Columbia University, N.Y.

[4] S. Anily and A. Federgruen. *Simulated annealing methods with general acceptance probabilities.* Technical Report, Graduate School of Business, Columbia University, New York, 1986. preprint, to appear in Journal of applied probability.

[5] B.A. Armstrong. A hybrid algorithm for reducing matrix bandwidth. *International journal for numerical methods in engineering*, 20:1929–1940, 1984.

[6] G. de Balbine. Note on random permutations. *Mathematics of Computation*, 21(100):710–712, 1967.

[7] P. Banerjee and M. Jones. A parallel simulated annealing algorithm for standard cell placement on a hypercube computer. In *Proceedings IEEE international conference on computer-aided design*, pages 34–37, IEEE, Santa Clara, November 1986.

[8] J. Beardwood, J.H. Halton, and J.M. Hammersley. The shortest path through many points. In *Proceedings Cambridge Philosophycal Society*, pages 299–327, 1959.

[9] J. Bernasconi. Low autocorrelation binary sequences: statistical mechanics and configuration space analysis. July 1986. presented at the IBM workshop on statistical physics in engineering and biology, Lech.

[10] S. Bernstein. Sur les fonctions absolument monotones. *Acta Mathematica*, 51:1–66, 1928.

[11] K. Binder. *Monte carlo methods in statistical physics*. Springer Verlag, New York, 1978.

[12] R. Biswas and D.R. Hamann. Simulated annealing of silicon atom clusters in langevin molecular dynamics. *Physical Review, B*, 34:895–901, 1986.

[13] I. Bohachevsky, M.E. Johnson, and M.L. Stein. Generalized simulated annealing for function optimization. *Technometrics*, 28:209–217, 1986.

[14] I. Bohachevsky, V.K. Viswanathan, and D.R. Rossbach. Generalized simulated annealing in the construction of "intelligent" design programs. July 1986. presented at the IBM workshop on statistical physics in engineering and biology, Lech.

[15] E. Bonomi and J.-L. Lutton. The asymptotic behaviour of quadratic sum assignment problems: a statistical mechanics approach. *European journal of operations research*, 26:295–300, 1986.

[16] E. Bonomi and J.-L. Lutton. The n-city travelling salesman problem: statistical mechanics and the metropolis algorithm. *SIAM review*, 26(4):551–568, October 1984.

[17] D.G. Bounds. Physics for travelling salesmen: some new approaches to combinatorial optimization. 1986. submitted to the bulletin of the Institute of Mathematics and its Applications.

[18] D. Braun, C. Sechen, and A.L. Sangiovanni-Vincentelli. A complete standard cell layout system. In *Proceedings 1986 custom IC conference*, pages 276–280, May 1986.

[19] R.G. Brown and R.F. Meyer. The fundamental theorem of exponential smoothing. *Operations research: the journal of the Operations Research Society of America*, 9:673–685, 1961.

[20] R.E. Burkard and F. Rendl. A thermodynamically motivated simulation procedure for combinatorial optimization problems. *European journal of operations research*, 17:169–174, 1984.

[21] P. Carnevalli, L. Coletti, and S. Paternello. Image processing by simulated annealing. *IBM journal of research and development*, 29:569–579, 1985.

[22] A. Casotto, F. Romeo, and A.L. Sangiovanni-Vincentelli. A parallel simulated annealing algorithm for the placement of macro-cells. In *Proceedings IEEE international conference on computer-aided design*, pages 30–33, Santa Clara, November 1986.

[23] F. Catthoor, H. DeMan, and J. Vanderwalle. Investigation of finite word-length effects in arbitrary digital filters using simulated annealing. In *Proceedings IEEE international conference on circuits and systems*, pages 1296–1297, San Jose, May 1986.

[24] F. Catthoor, H. DeMan, and J. Vanderwalle. Sailplane: a simulated annealing based cad-tool for the analysis of limit cycle behaviour. In *Proceedings IEEE international conference on computer design*, pages 244–247, Port Chester, October 1985.

[25] V. Černy. Solving inverse problems by simulated annealing. July 1986. presented at the IBM workshop on statistical physics in engineering and biology, Lech.

[26] V. Černy. Thermodynamical approach to the travelling salesman problem: an efficient simulation algorithm. *Journal of optimization theory and applications*, 45:41–51, 1985.

[27] V. Černy and I. Novàk. Picture processing by statistically coupled processors: relaxation syntactical analysis. July 1986. presented at the IBM workshop on statistical physics in engineering and biology, Lech.

[28] M.J. Chung and K.K. Rao. Parallel simulated annealing for partitioning and routing. In *Proceedings IEEE international conference on computer design*, pages 238–242, Port Chester, October 1986.

[29] D.P. Connors and P.R. Kumar. Simulated annealing and balance of recurrence order in time-inhomogeneous Markov chains. In *Proceedings 26th IEEE conference on decision and control*, pages 2261-2263, Los Angeles, 1983.

[30] F. Darema-Rogers, S. Kirkpatrick, and V.A. Norton. *Simulated annealing on shared memory parallel systems.* Research report RC 12195, IBM, 1986.

[31] S. Devadas and A.R. Newton. Genie: a generalized array optimizer for vlsi synthesis. In *Proceedings 23rd design automation conference*, pages 631–637, Las Vegas, June 1986.

[32] S. Devadas and A.R. Newton. Topological optimization of multiple level array logic: on uni and multi-processors. In *Proceedings IEEE international conference on computer-aided design*, pages 38–41, Santa Clara, November 1986.

[33] F. Distante and V. Piuri. Optimum behavioral test procedure for vlsi devices: a simulated annealing approach. In *Proceedings IEEE international conference on computer design*, pages 31–35, Port Chester, October 1986.

[34] P. Erdös, W. Feller, and H. Pollard. A theorem on power series. *Bulletin of the American Mathematical Society*, 55:201–204, 1949.

[35] D.A. D'Esopo. A note on forecasting by the exponential smoothing operator. *Operations research: the journal of the Operations Research Society of America*, 9:686–687, 1961.

[36] R. Ettelaie and M.A. Moore. Residual entropy and simulated annealing. *Journal de physique lettres*, 46:L–893 – L–900, 1985.

[37] U. Faigle and R. Schrader. *On the convergence of stationary distributions in simulated annealing algorithms.* Research report 86428, Institut für Operations Research, Bonn, W.Germany, October 1987.

[38] U. Faigle and R. Schrader. Simulated annealing - eine Fallstudie. Preprint WP 86430, Institut für Operations Research, Universität Bonn.

[39] W. Feller. *An introduction to probability theory and applications.* Volume 1, John Wiley & Sons, New York, London, Sydney, 3rd edition, 1968.

[40] R. Fiebrich and C. Wang. Circuit placement based on simulated annealing on a massively parallel computer. In *Proceedings IEEE conference on computer design*, Port Chester, 1987.

[41] G. Frobenius. Über Matrizen aus nicht negativen Elementen. *Sitzungs-Bereich Preußische Adademie der Wissenschaften*, 456–77, 1912.

[42] R.G. Gallager. *Information theory and reliable communication.* Wiley, New York, 1968.

[43] A. El Gamal and I. Shperling. Design of good codes via simulated annealing. April 1984. List of abstracts, presented at the IBM workshop on statistical physics in engineering and biology, Yorktown Heights.

[44] M.R. Garey and D.S. Johnson. *Computers and intractability: a guide to the theory of np-completeness.* W.H. Freeman and Company, San Francisco, 1979.

[45] R.C. Geary. Distribution of student's distribution for non-normal samples. *Journal of the Royal Statistical Society, Series B*, 3:178–184, 1936.

[46] S.B. Gelfand and S.K. Mitter. Analysis of simulated annealing for optimization. In *Proceedings of the 24th conference on decision and control*, pages 779–86, Ft.Lauderdale, December 1985.

[47] S. Geman and D. Geman. Stochastic relaxation, gibbs distribution, and the bayesian restoration of images. *IEEE transactions on pattern analysis and machine intelligence*, 6:721–741, 1984.

[48] J.W. Gibbs. *Elementary principles in statistical mechanics.* Yale University Press, New Haven, Conn., 1902.

[49] B. Gidas. Nonstationary Markov chains and convergence of the annealing algorithm. *Journal of statistical physics*, 39:73–131, 1985.

[50] B. Gidas. Global optimization via the langevin equation. In *Proceedings of the 24th conference on decision and control*, pages 774–78, Ft.Lauderdale, December 1985.

[51] L.P.P.P. van Ginneken and R.H.J.M. Otten. An inner loop criterion for simulated annealing. *Physics letters A*, 130:429–435, 1988.

[52] B.L. Golden and C.C. Sksiscim. Using simulated annealing to solve routing and local problems. *Naval research logistics quarterly*, 33:261–279, 1986.

[53] G. Gonsalves. Logic synthesis using simulated annealing. In *Proceedings IEEE international conference on computer design*, pages 561–564, Port Chester, October 1986.

[54] J.W. Greene and K.J. Supowit. Simulated annealing without rejected moves. In *IEEE international conference on computer design*, pages 658–663, Port Chester, November 1984.

[55] J.W. Greene and K.J. Supowit. Simulated annealing without rejected moves. *IEEE transactions on computer-aided design*, CAD-5:221–228, 1986.

[56] G.S. Grest, C.M. Soukoulis, and K. Levin. Cooling-rate dependence for the spin-glass ground-state energy: implications for optimization by simulated annealing. *Physical review letters*, 56:1148–1151, 1986.

[57] L.K. Grover. A new simulated annealing algorithm for standard cell placement. In *Proceedings IEEE international conference on computer-aided design*, pages 378–380, Santa Clara, November 1986.

[58] L.K. Grover. Simulated annealing using approximate calculations. 1978.

[59] L.K. Grover. Standard cell placement using simulated sintering. In *Proceedings 24th DAC*, pages 56–59, Miami, June 1986.

[60] B. Hajek. Cooling schedules for optimal annealing. *Mathematics of operations research*, 13(2):311–29, 1988.

[61] B. Hajek. A tutorial survey of theory and applications of simulated annealing. In *Proceedings 24th conference on decision and control*, pages 755–760, Ft. Lauderdale, December 1985.

[62] J. Hajnal. The ergodic properties of non-homogeneous finite markov chains. In *Proceedings Cambridge Philosophical Society*, pages 67–77, 1955.

[63] J. Hajnal. On products of non-negative matrices. In *Mathematical proceedings of the Cambridge Philosophical Society*, pages 521–30, 1979.

[64] J. Hajnal. Weak ergodicity in non-homogeneous markov chains. In *Proceedings Cambridge Philosophical Society*, pages 233–46, 1955.

[65] L.A. Hemachandra and V.K. Wei. Simulated annealing and error correcting codes. Bell Communications Research, Murray Hill, unpublished manuscript, 1984.

[66] J.L. van Hemmen and I. Morgenstern, editors. *Heidelberg colloquium on spin glasses.* Volume 192 of *Lecture notes on physics*, Springer Verlag, New York, Heidelberg, Berlin, 1983.

[67] R. Holley and D. Stroock. Simulated annealing via Sobolev inequalities *Communications in mathematical physics*, 115:553–569, 1988.

[68] T.M. Hsieh, H.W. Leong and C.L. Liu. Two-dimensional layout compaction by simulated annealing. In *Proceedings international symposium on circuits and systems*, pages 2439–2443, Helsinki, 1988.

[69] M.D. Huang, F. Romeo, and A.L. Sangiovanni-Vincentelli. An efficient general cooling schedule for simulated. In *Proceedings IEEE international conference on computer-aided design*, pages 381–3834, Santa Clara, November 1986.

[70] S. Hustin and A. Sangiovanni-Vincentelli. Tim, a new standard cell placement program based on the simulated annealing algorithm. May 1988. International workshop on placement and routing, Research Triangle Park, NC, USA.

[71] D. Isaacson and R. Madsen. *Markov chains.* Wiley, New York, 1976.

[72] D. Isaacson and R. Madsen. Positive columns for stochastic matrices. *Journal of applied probability*, 11:829–834, 1974.

[73] E.T. Jaynes. Prior probabilities. *IEEE transactions on systems science and cybernetics*, SSC-4(3):227–241, September 1968.

[74] D.W. Jepsen and C.D. Gelatt Jr. Macro placement by monte carlo annealing. In *Proceedings IEEE international conference on computer design*, pages 495–498, Port Chester, November 1983.

[75] D.S. Johnson, C.R. Aragon, L.A. McGeoch, and C. Schevon. Optimization by simulated annealing: an experimental evaluation. April 1984, revised version, 1986. List of abstracts, the IBM workshop on statistical physics in engineering and biology, Yorktown Heights.

[76] D.S. Johnson, C.R. Argon, L.A. McGeoch, and C. Schevon. Optimization by simulated annealing: an experimental evaluation. 1984. Preprint, AT & T Bell Laboratories, Murray Hill, N.Y.

[77] D.S. Johnson, C.H. Papadimitriou, and M. Yannakakis. How easy is local search? In *Proceedings annual symposium of foundations of computer science*, pages 39–42, Los Angeles, 1985.

[78] W. Kern. *On the depth of combinatorial optimization problems*. Technical report 86.33, Universität zu Köln, Köln, 1986.

[79] A. Khachaturyan. Statistical mechanics approach in minimizing a multivariable function. *Journal of mathematical physics*, 27:1834–1838, 1986.

[80] A.I. Khinchin. *Mathematical foundations of information theory*. Dover Publications, New York, 1957. translated by R.A. Silverman and M.D. Friedman.

[81] S. Kirkpatrick. Optimization by simulated annealing: quantitative studies. *Journal of statistical physics*, 34:975–986, 1984.

[82] S. Kirkpatrick, C.D. Gelatt Jr., and M.P. Vecchi. *Optimization by simulated annealing*. Research report RC 9355, IBM, 1982.

[83] S. Kirkpatrick, C.D. Gelatt Jr., and M.P. Vecchi. Optimization by simulated annealing. *Science*, 220:671–680, 1983.

[84] S. Kirkpatrick and G. Toulouse. Configuration space analysis of travelling salesman problems. *Journal de physique*, 46:1277–1292, 1985.

[85] A. Kolmogorov. Anfangsgründe der Theorie der Markoffschen ketten mit unendlich vielen möglichen Zuständen. *Matematičeskii Sbornik, N.S.*, 1:607–610, 1936.

[86] A. Kolmogorov. Zur Theorie der Markoffschen Ketten. *Mathematische Annalen*, 112:155–160, 1935.

[87] I. Kozniewska. Ergodicité et stationnarité des chaînes de markoff variables à un nombre fini d'états possibles. *Colloquium mathematicum*, 9:333–346, 1962.

[88] S.A. Kravitz. *Multiprocessor-based placement by simulated annealing*. Research report CMUCAD-86-6, SRC-CMU Center for Computer-Aided Design, Carnegie-Mellon University, Pittsburgh, February 1986.

[89] S.A. Kravitz and R. Rutenbar. Multiprocessor-based placement by simulated annealing. In *Proceedings 23rd design automation conference*, pages 567–573, Las Vegas, June 1986.

[90] S.A. Kravitz and R. Rutenbar. Placement by simulated annealing on a multiprocessor. *IEEE transactions on computer-aided design*, 6:534–549, 1987.

[91] P.J.M. van Laarhoven. *Theoretical and computational aspects of simulated annealing*. PhD thesis, Erasamus Universiteit, Rotterdam, 1987.

[92] J. Lam and J.-M. Delosme. Logic minimization using simulated annealing. In *Proceedings IEEE international conference on computer-aided design Santa Clara*, pages 348–351, Santa Clara, November 1986.

[93] H.W. Leong. A new algorithm for gate matrix layout. In *Proceedings IEEE international conference on computer-aided design*, pages 316–319, Santa Clara, November 1986.

[94] H.W. Leong, D.F. Wong, and C.L. Liu. A simulated-annealing channel router. In *Proceedings IEEE international conference on computer-aided design*, pages 226–229, Santa Clara, November 1985.

[95] P. Lévy. *Calcul des probabilités*. Gauthier-Villars, Paris, 1925.

[96] J.W. Lindeberg. Eine neue Herleitung des Exponentialgesetzes in der Wahrscheinlichkeitsrechnung. *Mathematische Zeitschrift*, 15:211–225, 1922.

[97] E. Lukacs. A characterization of the normal distribution. *Annals of mathematical statistics*, 13:91–93, 1942.

[98] M. Lundy. Applications of the annealing algorithm to combinatorial problems in statistics. *Biometrika*, 72:191–198, 1985.

[99] M. Lundy and A. Mees. Convergence of an annealing algorithm. *Mathematical programming*, 34:111–124, 1986.

[100] J.-L. Lutton and E. Bonomi. Simulated annealing algorithm for the minimum weighted perfect euclidean matching problem. *R.A.I.R.O. recherche opérationelle*, 20:177–197, 1986.

[101] A. Lyberatos, P. Wohlfarth, and R.W. Chantrell. Simulated annealing: an application in fine particle magnetism. *IEEE transactions on magnetics*, MAG-21:1277–1282, 1985.

[102] J. Masarik. A thermodynamically motivated optimization algorithm: circular wheel balance optimization. *Aplikace matematiky*, 20:413–423, 1985.

[103] N. Metropolis, A.W. Rosenbluth, M.N. Rosenbluth, A.H. Teller, and E. Teller. Equation of state calculations by fast computing machines. *Journal of chemical physics*, 21:1087–1092, 1953.

[104] M. Mézard. Spin glasses and optimization. In J.L. van Hemmen and I. Morgenstern, editors, *Heidelberg colloquium on glassy dynamics and optimization*, Springer Verlag, Berlin, 1987.

[105] M. Mézard and G. Parisi. A replica analysis of the travelling salesman problem. *Journal physique*, 47:1285–1296, 1986.

[106] M. Mézard and G. Parisi. Replicas and optimization. *Journal Physique Lettres*, 46:L–771 – L–778, 1985.

[107] M. Mézard, G. Parisi, N. Sourlas, G. Toulouse, and M. Virasoro. Nature of the spin-glass phase. *Physical review letters*, 52:1156–1159, 1984.

[108] D. Mitra, F. Romeo, and A.L. Sangiovanni-Vincentelli. Convergence and finite-time behavior of simulated annealing. In *Proceedings 24th conference on decision and control*, pages 761–767, Ft. Lauderdale, December 1985.

[109] T.P. Moore and A.J. de Geus. Simulated annealing controlled by a rule-based expert system. In *Proceedings IEEE international conference on computer-aided design*, pages 200–202, Santa Clara, November 1985.

[110] C.A. Morgenstern and H.D. Shapiro. *Chromatic number approximation using simulated annealing.* Technical report CS86-1, Department of Computer Science, The University of New Mexico, Albuquerque, 1986.

[111] I. Morgenstern. Phase transitions in spin glasses and chip design. In J.L. van Hemmen and I. Morgenstern, editors, *Heidelberg colloquium on glassy dynamics and optimization*, Springer Verlag, Berlin, 1987.

[112] R.C. Mosteller. *Monte carlo methods for 2-D compaction*, California Institute of Technology, Pasadena, 1986. Ph.D. dissertation.

[113] S. Nahar, S. Sahni, and E. Shragowitz. Experiments with simulated annealing. In *Proceedings 22nd design automation conference*, pages 748–752, Las Vegas, June 1985.

[114] S. Nahar, S. Sahni, and E. Shragowitz. Simulated annealing and combinatorial optimization. In *Proceedings 23rd design automation conference*, pages 293–299, Las Vegas, June 1986.

[115] H. Neudecker. Some theorems on matrix differentiation with special reference to Kronecker matrix products. *Journal of the American Statistical Association*, 64:953–963, 1969.

[116] W. Osman. Two-dimensional compaction of abstract layouts with statistical cooling. M.Sc. thesis, Philips Research Laboratories, Eindhoven, 1987.

[117] J. Nulton, P. Salamon, B. Andresen, and Qi Anmin. Quasistatic processes as step equilibrations. *Journal of chemical physics*, 83(1):334–38, July 1985.

[118] R.H.J.M. Otten and L.P.P.P. van Ginneken. *Annealing: the algorithm.* Research report RC 10861, Thomas J. Watson Research Center, IBM Corporation, Yorktown Heights, December 1984.

[119] R.H.J.M. Otten and L.P.P.P. van Ginneken. Floorplan design using simulated annealing. In *Proceedings IEEE international conference on computer-aided design*, pages 96–98, Santa Clara, November 1984.

[120] R.H.J.M. Otten and L.P.P.P. van Ginneken. Stop criteria in simulated annealing. In *Proceedings IEEE international conference on computer design*, pages 549–552, Rye Brook, October 1988.

[121] J.D. Pincus and A. Despain. Delay reduction using simulated annealing. In *Proceedings 23rd design automation conference*, pages 690–695, Las Vegas, June 1986.

[122] R.E. Randelman and G.S. Grest. N-city travelling salesman problem: optimization by simulated annealings. *Journal of statistycal physics*, 45:885–890, 1986.

[123] C.P. Ravikumar and L.M. Patnaik. Parallel placement based on simulated annealing. In *Proceedings IEEE international conference on computer design*, Port Chester, 1987.

[124] S. Rees and R.C. Ball. Criteria for an optimum simulated annealing schedule for problems of the travelling salesman type. *Journal of physics, A: mathematical and general*, 20:1239–49, 1987.

[125] F. Romeo and A.L. Sangiovanni-Vincentelli. Probabilistic hill climbing algorithms: properties and applications. In H. Fuchs, editor, *Proceedings 1985 Chapel Hill conference on vlsi*, pages 393–417, Computer Science Press, May 1985.

[126] F. Romeo, A.L. Sangiovanni-Vincentelli, and C. Sechen. Research on simulated annealing at Berkeley. In *Proceedings IEEE international conference on computer design*, pages 652–657, Port Chester, November 1984.

[127] J.S. Rose, D.R. Blythe, W.M. Snelgrove, and Z.G. Vranesic. Temperature measurement of simulated annealing placements. In *Proceedings ICCAD*, 1988.

[128] Y. Rossier, M. Troyon, and Th.M. Liebling. *Probabilistic exchange algorithms and euclidean travelling salesman problems.* Report RO 851125, EPF, Lausanne, 1985.

[129] D.E. Rutherford. In *Proceedings of the Koninklijke Nederlandse Adademie van Wetenschappen, Series A*, pages 54–59, 1932.

[130] S. Rothman. Circuit placement by simulated annealing at IBM. July 1986. presented at the IBM workshop on statistical physics in engineering and biology, Lech.

[131] R.A. Rutenbar and S.A. Kravitz. Layout by annealing in a parallel environment. In *Proceedings IEEE international conference on computer design*, pages 434–437, Port Chester, October 1986.

[132] P. Salamon, J.D. Nulton, and R.S. Berry. Length in statistical thermodynamics. *Journal of chemical physics*, 82(5):2433–36, March 1985.

[133] G.H. Sasaki and B. Hajek. The time complexity of maximum matching by simulated annealing. 1986. to appear in: Journal of the ACM.

[134] C. Sechen. *Placement and global routing of integrated circuits using the simulated annealing algorithm.* PhD thesis, University of California at Berkeley, Berkeley, 1986.

[135] C. Sechen and A.L. Sangiovanni-Vincentelli. The TimberWolf placement and routing package. In *Proceedings 1984 custom IC conference*, pages 522–527, Rochester, May 1984.

[136] C. Sechen and A.L. Sangiovanni-Vincentelli. The TimberWolf placement and routing package. *IEEE journal on solid state circuits*, SC-20:510–522, 1985.

[137] C. Sechen and A.L. Sangiovanni-Vincentelli. TimberWolf3.2: a new standard cell placement and global routing package. In *Proceedings 23rd design automation conference*, pages 432–439, Las Vegas, June 1986.

[138] E. Seneta. *Non-negative matrices and markov chains.* Springer Verlag, New York, Heidelberg, Berlin, 2nd edition, 1981.

[139] A. Sinclair and M. Jerrum. *Approximate counting, uniform generation and rapidly mixing markov chains.* Internal report CSR-241-87, University of Edinburgh, Department of Computer Science, Edinburgh, October 1987.

[140] C.C. Skiscim and B.L. Golden. Optimization by simulated annealing: a preliminary computational study for the TSP. 1983. presented at the N.I.H.E. summer school on combinatorial optimization, Dublin.

[141] W.E. Smith, H.H. Barrett, and R.G. Paxman. Reconstruction of objects from coded images by simulated annealing. *Optics letters*, 8:199–201, 1983.

[142] W.E. Smith, R.G. Paxman, and H.H. Barrett. Application of simulated annealing to coded-aperture design and tomographic reconstruction. *IEEE transactions on nuclear science*, NS-32:758–761, 1985.

[143] S. Solla, G. Sorkin, and S. White. Configuration space analysis for optimization problems. In E. Bienenstock, F. Fogelman, Soulié, and G. Weisbuch, editors, *Disordered systems and biological organization*, Springer Verlag, 1986.

[144] E.D. Sontag and H.J. Sussmann. Image restoration and segmentation using the annealing algorithm. In *Proceedings of the 24th conference on decision and control*, pages 768–73, Ft.Lauderdale, December 1985.

[145] P. Spira and C. Hage. Hardware acceleration of gate array layout. In *Proceedings 22nd design automation conference*, pages 359–366, Las Vegas, June 1985.

[146] H. Szu and R. Hartley. Fast simulated annealing with cauchy probability densities. 1986. Naval Research Laboratory, Washington DC, preprint.

[147] J.N. Tstsiklis. Markov chains with rare transitions and simulated annealing. To appear in: Mathematics of operations research.

[148] J.-P. Uhry. Simulated annealing for mesh building and other problems. July 1986. presented at the IBM workshop on statistical physics in engineering and biology, Lech.

[149] J.L. van Hemmen and I. Morgenstern, editors. *Heidelberg colloquium on spin glasses.* Volume 192 of *Lecture notes on physics*, Springer Verlag, New York, Heidelberg, Berlin, 1983.

[150] D. Vanderbilt and S.G. Louie. A monte carlo simulated annealing approach to optimization over continuous variables. *Journal of computational physics*, 36:259–271, 1984.

[151] J. Vannimenum and M. Mézard. On the statistical mechanics of optimization problems of the travelling salesman type. *Journal de physique lettres*, 45:L–1145 – L–1153, 19184.

[152] M.P. Vecchi and S. Kirkpatrick. Global wiring by simulated annealing. *IEEE transactions on computer-aided design*, CAD-2:215–222, 1983.

[153] S.R. White. Concepts of scale in simulated annealing. In *Proceedings of IEEE international conference on computer design*, pages 646–651, Port Chester, November 1984.

[154] L.T. Wille. The football pool problem for 6 matches: a new upper bound obtained by simulated annealing. 1986. to appear in: Journal of combinatorial theory, Series A.

[155] L.T. Wille. Searching potential energy surfaces by simulated annealing. *Nature: international journal of science*, 324:46–48, 1986.

[156] L.T. Wille and J. Vennik. Electrostatic energy minimisation by simulated annealing. *Journal of physics, A: mathematical and general*, 18:L1113–1117, 1985. Corrigendum, 19(1986)1983.

[157] L.T. Willle. Minimum energy configurations of atomic clusters: new results obtained by simulated annealing. 1986. to appear in: Chemical physics letters.

[158] D.F. Wong, H.W. Leong, and C.L. Liu. Multiple PLA folding by the method of simulated annealing. In *Proceedings of the 1986 custom IC conference*, pages 351–355, Rochester, May 1986.

[159] D.F. Wong and C.L. Liu. A new algorithm for floorplan design. In *Proceedings of the 23rd design automation conference*, pages 101–107, Las Vegas, June 1986.

[160] D.F. Wong and C.L. Liu. Floorplan design for rectangular and L-shaped modules. In *Proceedings IEEE international conference on computer-aided design*, pages 520–523, Santa Clara, 1987.

[161] D.F. Wong, H.W. Leong and C.L. Liu. *Simulated annealing for vlsi design.* Kluwer, Boston, 1988.

[162] W.W. Wood. *Monte carlo studies of simple liquid models*, pages 117–230. North Holland, Amsterdam, 1968.

INDEX

acceptance function, 9, 49, 79
  choice, 84
  constraints, 85
  probability, 20
accessibility, 120, 128
  chain, 77
  constraints, 86
  final, 138
  global, 141
  local, 78, 141
    estimate, 147
  maximal, 86
  properties, 75
  state space, 65
adjacent states, 48
aggregate
  function, 78
  logarithm of the score, 120
  three parameter, 122
algorithm
  performance, 9
algorithms, 4
annealing, 15
  algorithm, 11, 153
  chain, 77, 91
  convergence, 141
  curve, 135
  quality of solutions, 13
  schedule, 13
  simulation, 14
approximate calculations, 177
assignment problem, 1
average score, 74, 80
  hyperbola, 119
  linear behavior, 90
  linear, 119
average, 71

balance equation, 59, 89
barrier distance, 157
blind random search, 5
boltzmann density, 14
bounding box, 168

catalytic moves, 164
central limit theorem, 112, 129
chain convergence, 156
chain limit theorem, 57, 63, 142
chain, 49
  properties, 91
characteristic function, 96
  uniqueness, 105
characteristic

polynomial, 23
chi square distribution, 105
clipped state space, 157
cofactor, 22
column order, 21
combinatorial optimization, 1
tractability, 4
completing the square, 100, 109
complex state sequence, 81
computational efficiency, 164
computer, 4, 48
limited resources, 4
time, 4
conditional
density function, 67
entropy, 70, 77
probabilities, 67
conductance, 154
configuration, 1, 48
random, 5
congruent matrices, 42
constrained optimization, 87, 117
control parameter, 9, 49, 80, 127, 128
decrements, 131
convergence, 138
matrix expressions, 38
rate of, 147
covariance, 69
matrix, 69
crystal, 15
cycles of a permutation, 52
cyclic
sequences, 3
state sequence, 81
cylinder, 1

degrees of freedom, 14, 123
density of scores, 7
density of states, 115
detailed balance conditions, 59
determinant, 22
diameter of the state space, 48
bounds for, 52
dip, 157
discrete ensemble, 70
dispersion, 145
distances, 2
distribution function, 96
n-variate, 66
reproductive, 106
distribution of local minima, 7
drilling machine, 2
dyadic product, 22
dynamic aspect, 141

efficient, 4
eigenvalue, 23, 24
annealing chain, 143
diagonal, 27
kronecker product, 29
kronecker sum, 30
matrix norm, 38
multiplicity, 23, 32
real, 24, 63
second, 144, 150
spectrum, 23
stochastic matrix, 53
transition matrix, 53, 92
eigenvector
left, 24
right, 24
standardized, 24
energy, 14
average, 16
final, 16
entropy, 70, 73, 77, 117

joint, 70
equilibrium density function, 65
equilibrium density, 16, 47, 58, 79
equilibrium, 58, 141
equivalent quadratic forms, 42
Erdös, 63
ergodic, 92
process, 58
estimator, 71
exhaustive search, 161
expectation
linearity property, 72
expected values, 78

feasable configurations, 4
Feller, 63
frequency data, 71
Frobenius, 63

gamma density, 122
gamma
distribution, 104
function, 105
Gaussian distribution, 103
generation mechanism, 3
generation process, 5
Gershgorin's theorem, 42
global minimum, 160
global
convergence, 164
minimum, 7, 92, 138
graph partitioning, 18
ground states, 15

Hankel's contour integral, 104
hierarchy for local minima, 158
hill climbing algorithm, 9
hill climbing, 9
homogeneous markov process, 47

identity relation, 48
implementation, 167
incidence structure, 51
independent states, 144
inertia law of Sylvester, 46
information, 67
initialization, 128
inner loop criterion, 139, 141, 148
inner product, 22
integrated circuits, 18
integration theory, 113
interacting particles, 14
intermediate data structures, 164
invalid configurations, 128
invariant subspaces, 28
iterative improvement, 7

jacobian, 68
Jaynes' principle, 78
joint
density function, 66
distribution function, 66
entropy, 70
jordan form, 28

Kolmogorov, 63
conditions, 61
kronecker
product, 29
sum, 30

lagrange
multipliers, 74, 87
reduction method, 46
landscape, 9, 148, 156
laplace transform, 116
Lebesgue-measurable, 66
Lebesgue-Stieltjes integral, 66
Levy's theorem, 98

linear arrangement, 50
linear combination, 23
linear dependent, 23
linearity hypothesis, 124
linearly independent, 23
local minimum, 160
local
  minimum, 7, 157
    escape from, 9

marginal density function, 67
marginal distribution function, 67
markov chain, 19
  homogeneous, 19
markov process, 50, 77
  inhomogeneous, 148
markov property, 50
matrix, 21
  covariance, 69, 100, 103
  diagonal, 25, 26
  expressions
    convergence, 38
  identity, 22
  inverse, 23
  limit, 39
  lower triangular, 25, 26
  nonnegative, 64
  nonsingular, 23
  norm, 35
    consistent, 35
  partitioned, 28, 31
  positive definite, 45
  positive semidefinite, 45
  power, 25
  product, 25
  real, 21
  similar, 24
  singular, 23
  spectrum, 26
  square, 21, 23
  stochastic, 53
  subordinate
    norm, 36
  symmetric, 22
  theory, 21
  trace, 24
  transition, 52
  transpose, 22
  upper triangular, 25, 26
mean, 69
Metropolis, 14
minimum dependence period, 145
minkowski
  ∞-norm, 37
  1-norm, 37
  inequality, 35
  p-norms, 35
minor, 22
move distance, 157
move length, 175
move quality, 175
move set, 18
move size, 175
move, 6, 48
  constraints, 80
  set, 153, 168
multiple objectives, 3
mutual information, 68

neighbor relations, 48
neighborhood, 6
neudecker's identity, 31
non-trivial seminorm, 34
normal distribution, 103
  characteristic function, 101
  k-variate, 101

standardized form, 103
normal form of a matrix, 28
np-hard, 4
number of iterations, 146

objective function, 1
optimization problem, 1
optimization
unconstrained, 87
optimum state, 4
orthonormal eigenvectors, 25
outer loop, 13

permutations, 22
Perron-Frobenius, 63
physical analogue, 15
pin, 51
placement problem, 164
placement, 18
Pollard, 63
polynomial time, 4
positive definite, 45
positive semi-definite, 45, 72
printed circuit board, 2
prior information, 71
probability, 66
most likely law, 73
conditional, 67
density function, 66
theory, 65
problem instance, 4
pseudo-diagonal normal form, 32
pseudo-diagonal normal forms, 25

quadratic form, 42
signature, 43
quadratic forms, 103
quality
absolute measure, 3
indicator, 48
of an option, 3
sum of measures, 3
quasi-equilibrium, 15, 131

random state generation, 128
random
blind search, 5
configurations, 5
number, 9
problem instance, 163
process, 71
variable, 66, 96
absolutely continuous, 66
independent, 99
range limiting, 175
rate of convergence, 142, 162
reflexive chain, 54
reflexive move set, 91
rejectionless method, 176
relative state frequencies, 58
return sequence, 81
reversal, 163
reversible chain, 59, 61
rim level, 157
row order, 21

sample average, 107
sampling covariance matrix, 72
sampling distributions
asymptotic properties, 112
sampling theory, 96
scalar product, 22
schedule, 13, 127
score average, 128
score density, 58
score function, 3, 4, 18, 48, 65, 168
minimization, 3

score variance, 118, 128
second eigenvalue, 154
selection function, 65
  symmetry, 91
selection probability, 49
seminorm, 34
short-cut, 176
similar matrix, 24
similarity transformation, 26
simple state sequence, 81
size of a problem, 4
smoothing, 137
space conductance, 154, 156, 159
space connectivity, 91
space diameter, 160
space plots, 161
spectrum, 23
speed of convergence, 141
speed-up, 176
spheres, 14
standard deviation, 69
state encoding, 48
state equation, 14
state sequence, 81
state space, 18, 48, 167
  structure, 153
states, 4, 48
stationary points, 87
statistical mechanics, 14
statistically independent, 67
statistics, 65
Stephanos' rule, 29
Stirling's formula, 73, 124, 163
stochastic matrix, 53
stochastic process, 65
stop criterion, 10, 135
strong control, 119
structure of the state set, 48
superdiagonal, 28
swapping, 163
symmetric move set, 91

Taylor's theorem, 112
temperature, 14
topography of the state space, 157
topology of the state set, 48
trans-information, 70
transition matrix, 19, 52
  eigenvalue, 53
  eigenvalues, 92
  real eigenvalues, 63
transition probability, 19, 50, 141, 146, 148
transpose, 22
  properties, 22
  vector, 22
transposition, 3
traveling salesman problem, 2, 13
triangle inequality, 34

ultrametric space, 158, 159
unbiased estimator, 71
uncertainty measure, 74
updating the score function, 164

valley, 9, 158
variance, 69, 96
  sample, 107
  score, 74
vector, 21
  components, 22
  eigen-, 24
  length, 22
  linear combination, 23
  linear dependent, 23
  mean, 69

norm, 35
partitioned, 66
random, 66, 96
transpose, 22

weak control, 117, 119
wiring, 18